Pension Sustainability in China

Pension Sustainability in China: Fragmented Administration and Population Aging aims to investigate the impact of fragmentation and population ageing on pension sustainability in China. The book demonstrates how pension sustainability is compromised by various adverse effects produced by fragmentation, such as the moral hazard caused by the disarticulated intergovernmental fiscal responsibility. An overlapping generations (OLG) model is updated with the latest demographic data and is used to assess the impact of population ageing on pension sustainability. The book considers whether adjustment in retirement age can ensure long-term financial sustainability. It explores how, compared to the population ageing, the issues stemming from the fragmentation pose a more insidious threat to pension sustainability in China.

Randong Yuan is a research fellow at the Advanced Institute of Global and Contemporary China Studies, the Chinese University of Hong Kong, Shenzhen. He is also a fellow at the Institute for International Affairs, Qianhai.

International Population Studies

This book series provides an outlet for integrated and in-depth coverage of innovative research on population themes and techniques. International in scope, the books in the series cover topics such as migration and mobility, advanced population projection techniques, microsimulation modeling, life course analysis, demographic estimation methods and relationship statistics. The series includes research monographs, edited collections, advanced level textbooks and reference works on both methods and substantive topics. Key to the series is the presentation of knowledge founded on social science analysis of hard demographic facts based on censuses, surveys, vital and migration statistics. All books in the series are subject to review.

Population Change in Europe, the Middle-East and North Africa
Beyond the Demographic Divide
Edited by Koenraad Matthijs, Karel Neels, Christiane Timmerman, Jacques Haers and Sara Mels

Internal Migration
Geographical Perspectives and Processes
Edited by Darren P. Smith, Nissa Finney, Keith Halfacree and Nigel Walford

Europe's Population to 2050
Trends, Projections and Policy Scenarios
Edited by Philip Rees, Joop De Beer, Nicole Van der Gaag and Frank Heins

Syrian Refugees in Turkey
A Demographic Profile and Linked Social Challenges
Edited by Alanur Çavlin

Pension Sustainability in China
Fragmented Administration and Population Aging
Randong Yuan

For more information about this series, please visit: www.routledge.com/International-Population-Studies/book-series/ASHSER-1353

Pension Sustainability in China

Fragmented Administration and
Population Aging

Randong Yuan

Routledge
Taylor & Francis Group

LONDON AND NEW YORK

First published 2022
by Routledge
2 Park Square, Milton Park, Abingdon, Oxon OX14 4RN

and by Routledge
605 Third Avenue, New York, NY 10158

Routledge is an imprint of the Taylor & Francis Group, an informa business

© 2022 Randong Yuan

British Library Cataloguing-in-Publication Data
A catalogue record for this book is available from the British Library

Library of Congress Cataloging-in-Publication Data
Names: Randong, Yuan, author.
Title: Pension sustainability in China : fragmented administration and population ageing / Randong Yuan.
Description: Milton Park, Abingdon, Oxon ; New York, NY : Routledge, 2022.
| Includes bibliographical references and index.
Identifiers: LCCN 2021027863 (print) | LCCN 2021027864 (ebook) | ISBN 9781032022819 (hardback) | ISBN 9781032022826 (paperback) | ISBN 9781003182696 (ebook)
Subjects: LCSH: Pensions--China. | Population aging--China.
Classification: LCC HD7230 .R36 2022 (print) | LCC HD7230 (ebook) | DDC 331.25/20951--dc23
LC record available at https://lccn.loc.gov/2021027863
LC ebook record available at https://lccn.loc.gov/2021027864

ISBN: 978-1-032-02281-9 (hbk)
ISBN: 978-1-032-02282-6 (pbk)
ISBN: 978-1-003-18269-6 (ebk)

DOI: 10.4324/9781003182696

Typeset in Bembo
by SPi Technologies India Pvt Ltd (Straive)

Contents

Figures

Tables

Acronyms

ACFTU	All-China Federation of Trade Unions
BRS	Basic Residents' Scheme
CAF	Central Adjustment Fund
CPC	Communist Party of China
GDP	Gross Domestic Product
ILO	International Labour Organization
IPD	Implicit Pension Debt
LTSR	Long-Term Sustainable Replacement Rate
NDC	Notional Defined Contributions
OECD	Organisation for Economic Co-operation and Development
OLG	Overlapping Generations (Model)
PAYG	Pay-As-You-Go
PES	Public Employees' Scheme
PRC	People's Republic of China
UES	Urban Employees' Scheme
WTO	World Trade Organization

Acknowledgements

First and foremost, I would like to express my deep gratitude to my two supervisors at the University of Melbourne. Professor Christine Wong guided me as the primary supervisor throughout the four years of the PhD research, from which this book is a direct product. Not only did I benefit greatly from her expertise and insights in the research topic, the scholars and experts in China she introduced to me opened the door for the fieldwork opportunities and helped me to connect with a network of research collaborators, which is indispensable for this research and invaluable for my future career. I am also particularly grateful to my second supervisor Dr Tim Robinson for his great advice and patience in helping me improve the draft chapters. The conference and workshop opportunities he arranged in the first two years of the programme also gave me the chance to hone my analytical and presentation skills.

For the fieldwork in China, I would like to express my great appreciation to Professor Yu Jianxing, Professor Zhu Ling, Dr Lauren Johnston, Professor Yuan Cheng, Professor Lin Wanlong, Ms Yang Yilu, Professor Liu Xiaofeng, Dr Huang Biao, Ms Teng Hongyan, Dr He Dongni, Mr Ding Kai, Mr Zhao Fengwan, Ms Wang Hong, Ms Jiang Juan, Mr Li Chuanyu, Dr Jiang Liu and Mr Gui Shiqi, for their great support in providing interview opportunities with local officials and other participants across 12 provinces in China during the two rounds of fieldwork for this research in 2017 and 2019. I also extend my sincere thanks to every interviewee and participant of this research for providing the very valuable information and sharing their experiences in pension reform in China.

I would like to thank the University of Melbourne for providing the financial support for my studies through the Melbourne Research Scholarship. I also appreciate the funding from the Faculty of Arts, the Asia Institute and the Centre for Contemporary Chinese Studies in supporting me to conduct the fieldwork in China.

Finally, I would like to thank Ms Faye Leerink, Professor Philip Rees, Ms Nonita Saha, Ms Charlotte Taylor, Ms Claudia Austin and Ms Elizabeth Spicer at Routledge for their enthusiasm and professional advice, without which the timely publication of this book would not be possible.

Preface

With a rapidly ageing population and a highly fragmented pension system divided into over 2,000 pools managed separately by local governments, the financial sustainability of the Chinese pension system is facing serious challenges. This book aims to investigate the impact of population ageing and fragmented administration on pension sustainability in China. An overlapping generations (OLG) model is updated with the latest demographic data and used to perform a prospective assessment of the impact of population ageing on pension sustainability in China and to help determine whether adjustment in retirement age can ensure long-term financial sustainability under various demographic scenarios in the rest of the 21st century. After examining the history of pension reform and policy evolution in the context of overall development of China, the study conducts an analysis on the consequences of fragmentation based on both the evidence obtained from fieldwork and secondary data including policy documents and official statistics. The distortion in incentives for local governments is documented in case studies covering both the coastal and inland regions. These case studies demonstrate how pension sustainability is compromised by various adverse effects produced by fragmentation, such as the moral hazard caused by the disarticulated intergovernmental fiscal responsibility.

Overall, the findings of this research reveal that, compared to the population ageing, the issues stemming from the fragmentation pose a more insidious threat to pension sustainability in China. The retirement age reform alone can only provide a necessary but not sufficient condition for ensuring the system's long-run financial sustainability, abstracting from the significant negative impact of the fragmentation. Problems of moral hazard such as noncompliance by local governments and challenges of adverse selection resulting from the administrative loopholes in the highly decentralised system, if left unchecked, are classic reasons why insurance policies including pension schemes go bankrupt. Therefore, if China wants to ensure the long-term sustainability of the pension system, it is imperative to take its reform to the next level by defragmenting the system. The possibility of the fertility cliff and the danger of the de facto bankruptcy brought about by the population ageing further highlight the urgency to address the fragmentation as the underlying cause of the many defects of the system that are damaging pension sustainability.

1 Introduction

This chapter first introduces the background and the motivation for the study to be presented in this book on the sustainability of the pension system in China. Two salient features of China, namely (1) experiencing population ageing at unprecedented speed and scale and (2) having a highly fragmented pension system in terms of administration and financing, set the focus of this study and drive the formulation of the research questions, which are stated in the second section of the chapter. The third section describes the data and the methodology employed in the study. The last section gives an outline of the book.

1.1 Background and motivation

Since the Chinese government started to implement its decision to reform and open up the country in the late 1970s, China has undergone a momentous transformation boosted by over three decades of stable and fast growth. The economy registered an annual average growth rate of 9.8% in the gross domestic product (GDP) during the first 35 years of the still ongoing transition from a command economy towards a market-based one.[1] As a result, China has emerged as the second largest economy and the largest trading nation in the world (McCurry and Kollewe 2011; Bloomberg 2013), and it has become an increasingly important engine of growth to many other countries (Jenkins, Peters, and Moreira 2008; Tan, Abeysinghe, and Tan 2015). Thus, the prospect of China's future development not only matters for the benefits of its 1.4 billion people, but it also carries great impact for maintaining regional and global economic prosperity.

But the growth momentum of China has appeared to be tapering off in recent years in the aftermath of the 2008 global financial crisis. China's annual growth in GDP has seen a trend of slowing down, from 10.6% in 2010 to below 8% since 2012, at 7.4% in 2014, 6.9% in 2015 and 6.6% in 2018, the slowest since 1990.[2] The Chinese leaders have proactively accepted that China has already entered into the "New Normal Era", which should be characterised by slower but sustainable, high-quality growth for several decades to come, according to President Xi Jinping and Premier Li Keqiang.[3] However, if such slowing in GDP growth continues in the next five to ten years, to levels under 5% or even lower, China will face a tangible risk of falling into the middle-income trap, where the current "New Normal" could become "Abnormal" (Zheng 2015) and a stagnation in per capita

DOI: 10.4324/9781003182696-1

income may derail further economic development, which Mr Lou Jiwei, former Finance Minister of China, sees a greater-than-50-percent chance of happening (China Daily 2015).

With a rapidly ageing population (Cai 2010; Galor 2012) and an urbanising economy that is in many ways still underdeveloped (Tao and Xu 2007), there are serious challenges China faces and has to overcome if it wants to maintain a steadfast pace of development in order to avoid falling into the middle-income trap and eventually graduate as a socioeconomically advanced country. One of those challenges lies in the lack of a well-functioning pension system in China.[4]

A pension system prevents the individuals' myopic behaviours and their possible failures in recognising the financial needs in retirement (Diamond 1977), and it provides old-age income security for the individuals through consumption smoothing over the lifetime and insurance against the longevity risk (Barr and Diamond 2006). A compulsory pension programme enforced by the state overcomes the problems of adverse selection that can lead to the collapse of a private annuity market when individuals have substantial private information about their health and mortality (Hosseini 2015). The capability of a pension system administrated by the government to redistribute income also helps to address the market failures that can lead to a socially unacceptable level of income inequality and poverty among the elderly in a society (Mulligan and Sala-i-Martin 1999). For all these reasons, public pension systems have been introduced by the governments in many countries. By 1999, 167 countries around the world had set up some types of pension programmes (Zhigang Yuan et al. 2016). The public expenditures on pensions are often the largest item of welfare expenditures in the industrialised countries, accounting for 18% of total government spending on average among the OECD countries in 2013 (OECD 2017).

As a specific and important instrument in the purview of public finance, a properly designed pension system not only has direct implications for the welfare and wellbeing of hundreds of millions of Chinese people, but it could also serve as a key ingredient in the recipe for success of further reform. A successful reform of the pension system can be a pivot point for China to implement the necessary economic restructuring under the New Normal Era for the country to embark on a growth trajectory that could be noticeably slower than before but at the same time more sustainable, inclusive and stable. A linchpin in the nexus of employment, capital formation, productivity, labour incentives, demographics and urbanisation, pension reform has remained a top policy concern for the Chinese central government (Hussain 1994; Wang, Béland, and Zhang 2014b).

As mentioned above, the economic headwind China finds itself sailing into now is caused by many daunting challenges facing the country, which makes improving its pension system an important and urgent task. Two idiosyncrasies of China further accentuate the exigencies of the task. First, while population ageing is not unique to China, the country is and will be experiencing the demographic transition at unparalleled speed and scale (Mai, Peng, and Chen 2013; Chen et al. 2019), which puts in doubt the long-term financial sustainability of the system. Second, unlike any other country running a public pension programme, the Chinese system is managed by lower level governments often at county or city levels, resulting in over 2,000 separate pools of pension funds with different ground rules across the country (Zheng 2012), which makes the system complex and opaque and complicates the policy implementation process. Taking these two factors into consideration, this study sets as its topic the pension reform in China. By focusing the study on the effects of the two factors on the financial sustainability of the pension system, the study presented in this book investigates the recent reform of the system in pursuit of a better system conducive to the long-term stability and prosperity of China.

1.2 Research questions

To investigate the recent reform of the pension system in China and to address the underlying research problem of how to further improve the current system and help ensure its long-term financial sustainability, the research questions for this study are formulated as follows.

1   How does the population ageing influence the financial sustainability of the pension system in China? Specifically,

    a) What impact will the fertility trends of China generate on the sustainability of its pension system under the plausible reform options in retirement age?

    b) How do different plausible reform options in financing method, retirement age and timing of implementation affect overall welfare outcome across generations while ensuring financial sustainability under different fertility scenarios?

    c) Is there a fertility cliff under which the pension system will be in de facto bankruptcy with all plausible retirement age reform options?

2   How does the fragmentation in pension administration and financing influence the financial sustainability of the pension system in China?

The impact of the population ageing on pension sustainability in China has been extensively discussed in the literature (West 1999; Wang 2002; Yuan and Wan 2006; Jin 2010; Liu 2013; Yuan 2013; Yuan 2014; Yu and Zeng 2015; Song et al. 2015). The debate is not settled, however, on whether retirement age reform can fundamentally solve the funding crisis facing the pension system in China. The existing literature also tends to tackle the problem by making rather optimistic fertility assumptions without considering the possibility of a fertility cliff, under which the pension system will be in de facto bankruptcy with all plausible retirement age reform options. Any pension system can be made sustainable in the nominal sense by adjusting internal system parameters such as contribution rate, benefits level, retirement age or a combination of those. But a pension system can be considered in de facto bankruptcy if the sustainability of the system can only be ensured by implementing an unrealistically high contribution rate, or a miserably low benefits level, or an exorbitantly high retirement age. Thus, the sustainability of the Chinese pension system can be assessed in terms of whether the de facto bankruptcy can be prevented, which not only depends on the internal system parameters but also is contingent on whether such a fertility cliff exists and, if so, where China is now in relation to the fertility cliff. This study attempts to enrich the current debate by investigating the impact of population ageing on pension sustainability in China based on the notion of de facto bankruptcy as a more nuanced criterion for determining whether the long-term sustainability of the system can be ensured under a wider range of fertility scenarios to take into account the possibility of a fertility cliff.

Many researchers have made scholarly contributions to understanding the complexity of the pension system and its reform in China in the past few decades (Barkan 1990; d'Haene and Emile 1994; Hussain 1994; Hu 1997; West 1999; Zhu 2002; Dong and Ye 2003; Leung 2003; Salditt, Whiteford, and Adema 2007; Williamson, Price, and Shen 2012; Oksanen 2012; B. Zheng 2012; Wu 2013; Wang, Béland, and Zhang 2014a; Cai and Cheng 2014; Zhao, Wang, and Mi 2015; Chen and Turner 2015; Zhu and Walker 2018). Various issues of pension reform in China have been identified and analysed in different aspects of the pension system such as its coverage, adequacy, fairness and financing method. However, the institutional feature of the Chinese pension system being highly fragmented and decentralised and its potential influence on pension sustainability have not attracted enough attention in the existing literature. The availability of officially released data on pension administration and financing is limited at local level,[5] which may have contributed to the lack of in-depth analysis on the implications of the fragmentation on pension sustainability in the scholarship to date. This project attempts to fill this gap by studying such unique institutional feature of the system and its effects on pension sustainability in China. The limited availability of secondary data on the pension system at local level highlights the importance of producing primary data for investigating how the fragmentation affected the operation of the system in various local contexts. The data collected from two rounds of fieldwork conducted for this project in 2017 and 2019 in China thus form an important part of the empirical evidence for the study.

While China has its idiosyncrasies in terms of the pace of ageing of its population and the administrative and financial structure of its pension system, it is by no means the only country facing the problem of "growing old before getting rich". Several countries, such as Romania and Ukraine, have already been facing similar challenges brought about by the problem of a society entering an advanced stage of population ageing before turning into a high-income economy (Johnston 2019). More developing countries, including those in East Asia and South America in particular, are expected to meet the mega-trend of ageing and set to join this "old before rich" group (Runde 2020). Many countries, such as most OECD countries, tried delaying retirement age to alleviate the financial burdens of the pension system caused by population ageing (OECD 2013). This study will try to explore this option and ascertain if retirement age reform alone is sufficient for ensuring pension sustainability in China's context. Other countries, including Poland and Russia in Europe as well as Argentina and Chile in South America, sought to solve the problem by moving away from pay-as-you-go (PAYG) towards funded systems (Montes and Riesco 2018); whether this constitutes a viable option for China will also be studied in this research. The "old before rich" problem is not necessarily an insurmountable one, as the experiences of quite a number of countries show, a developing country can indeed become rich after getting old (Johnston 2019). But to realise that entails solving a series of problems associated with the "old before rich" situation. Ensuring pension sustainability is one of them. As such, the lessons learnt from China's recent pension reform can be useful input for policy makers of other developing countries that are either already "old but not yet rich" or expected to confront similar issues in the next few decades. Thus, as an effort for gaining a better understanding of the problems China faces in pension sustainability, this research may also contribute to alleviating such a global issue.

1.3 Data and methodology

Both qualitative and quantitative analyses were used to seek answers to the research questions described above. All relevant and publicly available documents were within the scope for the qualitative analysis. The sources of the documents included directives from the Communist Party of China, legislations by the National People's Congress, policy documents released by the central government such as work reports of the State Council, policy documents released by relevant ministries including the Ministry of Human Resources and Social Security and the Ministry of Finance, as well as policy documents released by various levels of governments in China. Other sources such as academic reports and media coverage were also utilised as input whenever relevant.

To answer the first research question, a dynamic macroeconomic model with overlapping generations of forward-looking agents developed for modelling the Chinese economy (Song et al. 2015) was updated with the latest demographic data and used to perform a prospective assessment of the impact of the demographic trends in China on the pension system and help determine whether adjustment in retirement age can ensure the long-term fiscal sustainability of the system under different plausible scenarios concerning the demographic trends of China in the rest of the 21st century.

The primary data generated from the fieldwork carried out for this project form an important source of evidence for the empirical study for addressing the second research question concerning the impact of the fragmentation on pension sustainability. During two rounds of fieldwork conducted in 2017 and 2019, I visited 12 provinces and carried out semi-structured interviews with 35 local government officials in charge of the running of the local pension system in 21 counties or cities, covering all four regions of China consisting of the eastern costal area, central China, western China and north-eastern China.[6]

The selection of fieldwork sites was determined by the availability of interview opportunities with local officials, which were difficult to acquire without the help of local contacts who had close connections with the interview targets. I made several attempts in June 2017 to obtain interview opportunities with local officials through "cold calling" the government departments, but my requests were all promptly declined. All interviews with local officials that were successfully conducted for this study were arranged by the local contacts.[7]

The semi-structured interviews with local officials were designed to explore in detail the issues of local implementation of pension policies and conducted by following the interview guide described in Appendix A. When the interview process progressed, new and relevant information were gradually discovered and used to inform the subsequent interviews during the fieldwork. Consequently, new and more specific questions were added to modify the interviews heuristically. Such evolution of interview questions is reflected in the differences between the two versions of the interview guide used for the two rounds of fieldwork conducted in 2017 and 2019, respectively.

To complement the information acquired from the interviews with the local officials, I visited and had discussion with nine scholars specialising on pension policy studies in a few universities and research institutes in China during the fieldwork in 2017. During both rounds of fieldwork, I also conducted informal interviews with 48 migrant workers and 21 local residents randomly selected in the areas visited and discussed the relevant issues with three entrepreneurs, a bank manager and a job agency employee who had personal experience with the pension system. All interviews conducted during the fieldwork were transcribed with any identifying information about the interviewees anonymised. The data obtained from fieldwork interviews were analysed using qualitative data analysis methods including coding and thematic analysis. Case studies were used to report some key findings from the analysis of interview data, which help illustrate the behavioural patterns of local governments and their effects on pension sustainability.

By combining the qualitative and quantitative analyses, the current pension system in China was comprehensively studied and evaluated to gain a bettering understanding of how the population ageing and the fragmented structure of the pension system affect the financial sustainability of the system. Policy implications were then drawn based on the results of the study.

1.4 Outline of the book

The very fragmented nature of the Chinese pension system, which is only a part and an example of the highly fragmented and decentralised public finance regime in China (Wong 2007b) and a symptom of the country's lopsided intergovernmental fiscal system (Wong 2009, 2012), is a prominent candidate for the root cause of many problems of the system such as its inefficiency and lack of transparency, portability and fairness, which can affect pension sustainability directly or indirectly. Ultimately, the long-term sustainability of a pension system depends on a robust economy that can maintain a steady pace of development. The causal nexus runs both ways, as policy decisions made in the pension reform have an active impact on many factors that collectively determine long-term economic growth.

The rest of the book is organised as follows. Chapter 2 provides a brief historical account of the development of pension polices in China and a description of the current system. Chapter 3 reports on the quantitative study of the impact of population ageing and the effects of retirement age reform on the financial sustainability of Chinese pension system. Chapter 4 studies the impact of the fragmentation on pension sustainability by analysing the moral hazard present in the system with case studies containing the empirical evidence for the distortion in incentives for the local governments; the chapter also investigates the inefficiencies caused by the fragmentation that have direct negative impact on the financial sustainability of the system. Chapter 5 examines another mechanism through which the fragmentation can affect pension sustainability in China, namely the unfair treatment of migrant workers. Chapter 6 concludes the book by summarising the main findings of this research and drawing some policy implications on pension reform in China based on the findings.

Notes

1 The annual average growth rate of GDP for China for the period from 1978 to 2013 is calculated using data from the World Bank (World Bank 2018b).
2 China's GDP growth rates from 1990 to 2018 are based on the data from the World Bank (World Bank 2020a).
3 The term "New Normal Era" used for describing the current economic conditions of China was initially introduced by Xi Jinping in his speech during the Asia-Pacific Economic Cooperation (APEC) CEO Summit held in Beijing in November 2014 (Xinhua News Agency 2014) and further elaborated by Li Keqiang in his Report on the Work of the Government delivered in March 2015 (K. Li 2015).
4 Lou Jiwei listed pension reform as one of the four necessary ingredients for China to avoid falling into the middle-income trap, with the other three being agriculture reform, household registration system reform, and ensuring inclusive urbanisation and labour mobility (Lou 2015).
5 Availability of data on the pension system varies across the regions, making cross comparison difficult. Even at a relatively high level of the government such as the provincial level, basic financial figures such as total pension contributions cannot be obtained for many provinces in years as recent as 2017.
6 The 12 provinces visited for the fieldwork are Beijing, Heilongjiang, Shandong, Hebei, Zhejiang, Shanghai, Fujian, Guangdong, Sichuan, Hunan, Anhui and Guangxi.
7 Only in two occasions (in Shandong in 2017 and Anhui in 2019) did interviews fail to proceed even with the introduction by the local contacts, when the local officials claimed that there had been a misunderstanding and believed that I had the official introduction letter issued by their counterpart higher-level government departments, which I did not. Local contacts were obtained through personal connections.

2 Historical background and overview of the current system

This chapter lays out the historical background of pension reform in China with a focus on legacies and issues related to its sustainability. The first section provides an account of the development of the pension system in China since its inception to trace how it has evolved into the current form. Five stages of the pension reform in China, spanning from the early 1950s until the late 2010s, are discussed in detail. The gradualist, incremental, fragmentary and inadequate style of the reform is highlighted as a common characteristic of the pension reform measures explored in the chapter. The second section presents the components and key features of the current system.

2.1 Evolution of the pension system in China

To evaluate the current pension system in China and to study its sustainability, it is essential to have an understanding of the system anchored in a historical context. The evolution of the pension system in China can be juxtaposed with the socio-economic development of the country since the founding of the People's Republic of China (PRC). Five distinctive phases of the development of the Chinese pension system can be identified, and they are discussed in this section of the chapter.

2.1.1 Inception and early years of the pension system in China (1949–1965)

Public pension did not exist before the founding of the PRC; the elderly that became unable to earn an income then primarily relied on informal forms of financial support without the intervention of the state, which normally occurred within families (Zhigang Yuan et al. 2016).

On 26 February 1951, the "Labour Insurance Regulations of the PRC" was promulgated by the central government, signifying the beginning of the pension system in China. The pension system established by this policy document only covered state-owned, collective and private enterprises with more than 100 employees, which were geographically concentrated in the urban areas. In terms of financing of the system, the enterprises shouldered most of the financial responsibilities with some support by the All-China Federation of Trade Unions (ACFTU), while the employees did not need to pay any contributions (Li 2000). The system provided limited levels of pension benefits, with the replacement ratio ranging

DOI: 10.4324/9781003182696-2

from 35% to 60%. A male employee with at least 25 years of total employment history and at least ten years of employment history with the last enterprise he worked for was entitled to retirement benefits. For a female employee to be entitled to retirement benefits, the minimum total employment history and the minimum employment history in the last enterprise she worked for were 20 years and 10 years, respectively. The stipulated retirement age was 60 for male employees and 50 for female employees, with the flexibility of allowing an employee older than the retirement age to continue working with the consent of both the enterprise and the employee. Those employees still working after the retirement age were paid an extra allowance of 10% to 20% of their wages on top of the normal wages (State Council 1951).

Enterprises participating in the labour insurance were required to make monthly payment equivalent to 3% of the total salaries for their employees. For each participating enterprise, the first two monthly payments since the implementation of the labour insurance would be made to the ACFTU to build up a national labour insurance fund. From the third monthly payment onwards, 30% of these monthly payments made by each participating enterprise would be channelled into the national labour insurance fund managed by the ACFTU; the remaining 70% would be deposited into the enterprise-specific labour insurance fund managed by the trade union affiliated under each enterprise. Each participating enterprise would draw from its own labour insurance fund to pay for the pension benefits for its retirees; in each month, any surplus remaining in the labour insurance fund of each enterprise would be channelled into an account managed by the trade union at provincial or prefectural level or managed by the industry-specific trade union at the national level, in order to build up a labour insurance coordination fund. The labour insurance coordination funds would be used by the provincial or prefectural level trade unions or industry trade unions at the national level for the purpose of subsidising enterprises under their purview with shortage in the enterprise-specific labour insurance funds so as to honour the pension benefits due. The ACFTU was empowered to control and transfer funds across the labour insurance coordination funds. The trade unions managing the labour insurance coordination funds would apply to the ACFTU for subsidies if there was any deficit in these labour insurance coordination funds (State Council 1951). Such arrangements for financing the pension benefits for enterprises retirees amounted to a PAYG system with minimal accumulation of funds and limited risk pooling.

Some changes were made to the newly established pension system within two years, when the Ministry of Labour issued the "Draft Amendments to the Rules for Implementation of Labour Insurance Regulations of the PRC" on 26 January 1953. The retirement benefits were improved substantially, with the increased replacement ratio now ranging from 50% to 70%. The minimum employment history in the last enterprise an employee worked in was reduced to five years for both male and female employees (Ministry of Labour 1953).

The State Council issued the "Interim Measures for Retirement of Staff of State Organs" on 29 December 1955, which was the first official document to set out the rules over retirement and pension policies for civil servants and public institutions employees. As a result, a separate pension system was created for civil servants and public institutions employees, running independently from the pension system for employees in enterprises. The pension system for civil servants and public institutions employees was funded through public finance, had the same rules over retirement age as per the pension system for employees in enterprises, but provided slightly more generous pension benefits with the replacement ratio ranging from 50% to 80% (State Council 1955).

But the separation of pension rules between enterprises employees and civil servants and public institutions employees did not last long. Within two years, the Ministry of Labour and the ACFTU jointly drafted a policy document to merge the two pension systems, which was passed by the National People's Congress and the State Council as the "Provisional Regulations on the Retirement of Workers and Staff by the State Council", released on 9 February 1958. Male employees and male civil servants could retire at 60 if they had worked consecutively for at least five years and had a total employment history of at least 20 years. Female employees in managerial roles and female civil servants could retire at 55, while female employees in nonmanagerial roles could retire at 50, if they had worked consecutively for at least five years and had a total employment history of at least 15 years. The replacement ratio ranged between 40% and 70%, depending on factors such as employment history (State Council 1958). It is worth noting that the rules over retirement age of the Chinese pension system have essentially stayed the same since 1958, despite that many a sea change has occurred in the meanwhile in the socioeconomic and demographic structure of the country.

Although the same set of rules governing retirement entitlement and benefits was then applicable for both enterprises employees and civil servants and public institutions employees as prescribed by the 1958 policy provisions, the financing of the pension benefits was still organised separately between the two groups of employees. The financing of the pension benefits for enterprises employees was largely based on individual enterprises with a very limited level of risk pooling and coordination of funds by the ACFTU and its subsidiaries at provincial, prefectural or industry level. The financing of the pension benefits for civil servants and public institutions employees was paid through public finance at county or district level (State Council 1958). For both groups of employees, the financing of the pension system was highly fragmented since its inception. The fragmented financing would translate, in this case, into the fragmented and decentralised provision of pension services, due to the nature of the pension system being a cash collection and disbursement system. Until this date, such fragmentation and decentralisation remain as a salient feature of the Chinese pension system that makes it a rather peculiar case among the pension systems adopted by various countries in the world, as the responsibilities of providing vital basic public services such as social security are delegated to the local governments (Wong 2007b).

The newly established pension system played an important role in boosting the morale of the labour force and rehabilitating the economy of the war-torn country during the early years of the PRC. However, the rural residents were not included under the protection of the pension system when it was first conceived and established, and they remained out of the scope of the pension system during this first phase of the development of pension policies in China.[1] Moreover, a dichotomy between enterprises employees and public sectors employees was also engrained in its birth, and the duality in the financing mechanism was retained even after an apparent merge of the pension rules for the two groups of employees. Most unfortunately, the already weak element of risk pooling of the pension system was to be made worse during the next phase of its development.

2.1.2 Disruption to the pension system during the Cultural Revolution (1966–1976)

After delivering a reasonably stable performance in serving its purpose of providing pension coverage for employees from enterprises and public sector work units for more than a decade, the Chinese pension system was dismantled during the ten years of the Cultural Revolution from 1966 to 1976. The ACFTU was dissolved, leaving a vacuum in the role of coordinating and monitoring the pensions system. The pension funds accumulated until 1966 were appropriated for other uses, which effectively ended any risk pooling of the pension system built up prior to the Cultural Revolution (World Bank 1997). The Ministry of Finance issued the "Opinions on the Reform of Several Rules in the Finance of the State-Owned Enterprises (Draft)" in February 1969, which required the state-owned enterprises to cease the participation in the pension system as established during the 1950s by stopping making the monthly contributions to the labour insurance funds and to list pension benefits for their retirees as an extra item of operating costs (Ministry of Finance 1969). As the state-owned enterprises had become a major employer in urban parts of China by this time, the document essentially signified the dissolution of the pension system.

While employees were still exempted from making contributions, the financial responsibility to provide pension benefits for retirees fell back on individual enterprises or work units (Dixon 1981). The ability to fund the pension benefits was then largely determined by the financial strength of the individual enterprises or work units and the distribution of young and elderly employees in each enterprise or work unit; some of those enterprises facing financial difficulties were even unable to make timely payment for the pension benefits (Xu and Wan 2008). Due to the disruption to the pension system, many elderly employees who were supposedly eligible for retirement were not allowed to retire and "kept on working as long as physically capable" during the Cultural Revolution (Hussain 2007). Even two years after the end of the Cultural Revolution, in 1978, over two million elderly employees from enterprises and over 600 thousand elderly employees from public institutions and government agencies were still waiting to retire, even though they all exceeded the statutory retirement age (Ministry of Human Resources and Social Security 2011b). The pension system in its previous form that was coordinated by the ACFTU and its branches across the country no longer existed, resulting in a decreased level of risk pooling and income redistribution (Ministry of Human Resources and Social Security 2011b).

2.1.3 Small steps of pension reform in the era of fiscal decline (1977–1996)

Two years after the end of the Cultural Revolution, the central government decided to start the economic reform and opening up. Held from 18 to 22 December in 1978, the 3rd Plenary Session of the 11th Central Committee of the Communist Party of China signified the beginning of China's transformation from a planned economy to a market-based one. By gradually removing state control on prices and allowing more market competition, the economic reform led to a deterioration of profitability in the state-owned enterprises. As the taxes from the state-owned enterprises constituted a major source of government revenues, this contributed to a fiscal decline as an immediate effect of the market reform (Wong and Bird 2005). Both the share of the government revenues in the economy and the central government share of revenues fell sharply in the late 1980s and the early 1990s. Starting from 1978, such fiscal decline spanned for almost two decades. The size of the budget as measured by the ratios of both government revenues and expenditures to GDP only started to rise two years after the tax sharing system (fenshuizhi) reform, which was implemented in 1994 with the clear objective of stemming the issue of fiscal decline and some major efforts to specifically increase the share of central government revenues (Wong 2013b).

Against the backdrop of the prolonged fiscal decline, various fiscal reform initiatives were taken in a sequential manner, such as the tax for profits reform in 1984, the fiscal responsibility system reform in 1988 and the tax sharing system reform in 1994. These attempts to adjust the fiscal system for fixing the problems caused by the fiscal decline, however, had occurred without the guidance of any overall strategy to address the fundamental cause of the problems; the only discernible underlying logic seems to be that the decision to take each step of the fiscal and taxation reform was often made to react to the unwanted consequences introduced by the previous round of incremental reforms (Wong 1992). As a specific area of the general reform of the fiscal system in China, the pension reform in this period was carried out in a similarly reactive, piecemeal manner. As discussed below in this subsection, a series of small reform steps was taken to make adjustment to the existing pension system, often necessitated by reality rather than aiming to revamp it with a clear top-level design to achieve an ideal pension system conducive to the long-term prosperity and stability of the economy and society.

The pension system in China received some policy adjustment within two years after the end of the Cultural Revolution in 1976. The State Council issued the "State Council Circular on Issuing 'State Council Temporary Measures on Providing for Old, Weak, Sick and Handicapped Cadres' and 'State Council Temporary Measures on Workers' Retirement and Resignation'" on 2 June 1978. The policy document prescribed two measures for making the system more generous. First, the pension benefits level was improved by raising the replacement ratio to the range between 60% and 90% depending on the employment history. Second, the minimum employment history for pension entitlement was reduced to ten years for both male and female employees (State Council 1978). No other significant changes were made and, most critically, the financing of the pension system remained enterprise based.

Despite the lack of risk pooling, the enterprise-based pension system operating in the early stage of the economic reform was still largely financially viable even with increased burdens caused by improved pension benefits level, due to the fact that the number of retirees was low relative to the number of employees in most enterprises. The pension dependency ratio, i.e. the ratio of employees to retirees, was estimated to be around 30 to 1 in state-owned enterprises and collectively owned enterprises by the end of 1970s when the economic reform started (National Bureau of Statistics 1997).

The financing of the pension system remained highly fragmented with each enterprise or work unit operating its own pension fund. And many urban employees, especially those working in smaller enterprises that were not state owned, remained outside of the system with their needs for pension protection unmet. For the system to live up to the promise of becoming an important tool of public finance to help stabilise and promote the much needed economic growth in China, more significant reform measures had to be taken.

But instead of taking any bold moves to shore up the fragmented pension system, the central government was slow in its decision to incur any substantial changes to the existing system. Various reform proposals were put under test, as can be seen in a series of pilot reform programmes on the pension system carried out in different provinces in this period. Among them, one reform experiment made the breakthrough to turn the enterprise-based pension system into one that was based on social insurance locally managed at county or city level, which was conducted in several cities and counties in Guangdong, Sichuan, Jiangsu and Liaoning in 1984 (Ministry of Human Resources and Social Security 2011b).

Such local reform initiatives to increase the level of risk pooling for the pension funds were endorsed by the central government two years later, when the State Council issued the "Circular of the State Council on Issuing the Four Provisions for the Reform of the Labour System" on 12 July 1986. But instead of installing a pension system based on social insurance to cover all the employees, the central government was cautious and only took a small step in reforming the pension system. According to the policy document, only contract employees in state-owned enterprises would participate in the pension plan based on social insurance, with enterprises paying 15% of the payroll and contract employees paying up to 3% of their salaries to the local authority in charge of social insurance; permanent employees in the same enterprises were not subject to the reform (State Council 1986).

Within five years, the State Council issued the "Decision on the Reform of the Old-Age Insurance System for the Employees of Enterprises" on 26 June 1991. The policy document set out the rules for a pension system applicable to all employees in state-owned enterprises and allowed other types of enterprises to participate under the same system. The enterprises were required to pay a certain percentage of the payroll to the pension fund pooled at county or prefecture level. The provincial level government would determine the contribution rate for enterprises following the principle that the total contributions would be sufficient to cover the total payouts of pension benefits with a moderate surplus each year. The employees were also required to pay a certain percentage of their salaries, which would start at a rate below 3% and be later adjusted higher with wage increase and economic development at the discretion of the local governments (State Council 1991).

More pension reform experiments would be conducted in various parts of the country in the following years. As would be borne out later, different ideas for pension reform were to originate from different regions and different line ministries. And the policies proposed by them were to compete against each other. These policies proposed by the provinces or line ministries could be quite different as they were in turn dictated by the large socioeconomic disparities across different regions of the country, as in the case of the provinces, or motivated by the inherent difference in mandate, as in the case of the line ministries. Sometimes one of the ideas would win. More often a compromise would be reached so that a mix of the competing policies would be adopted as the national policy. In any case, the incremental nature of the reform initiatives was never changed.

As the economic reform was gathering momentum through the 1980s and 1990s, a demographic undercurrent was developing into a formidable force that could wreak havoc on the financial viability of the Chinese pension system in the foreseeable future. The high fertility rates that persisted during the first 20 years after 1949 declined drastically from 5.7 births per woman in 1970 to 2.8 in 1979 (World Bank 2018a). The one-child policy activated in 1979, combined with other factors such as economic development and urbanisation, also contributed to the further reduction of fertility to below replacement level since 1992 (World Bank 2021). Meanwhile, with mortality rates falling steadily since the founding of the PRC, life expectancy increased from 43.8 years in 1960 to 70.2 years in 1995 (World Bank 2018c). The combined effects of fertility decline and life expectancy increase in China inevitably resulted in population ageing, as the proportion of the elderly had increased substantially and will continue to increase in the coming decades with a falling ratio of employees to retirees. As shown in Figure 2.1, the

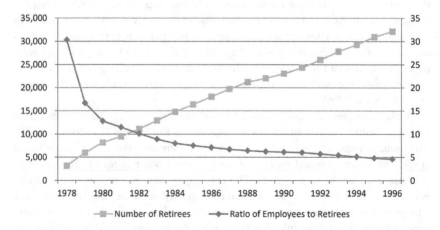

Figure 2.1 Number of retirees (in thousand) and ratio of employees to retirees, 1978–1996.

Sources: China Labour Statistical Yearbook 1997 (National Bureau of Statistics 1997).

ratio of employees to retirees was as high as 30.3 in 1978 in state-owned and collectively owned enterprises, but the same ratio dropped dramatically in one decade to 6.4 in 1988 and further declined to 4.6 in 1996; the number of retirees increased by 10 fold during this period, from 3.1 million in 1978 to 32.1 million in 1996 (National Bureau of Statistics 1997). With the worrying demographic trends, the long-term financial sustainability of the pension system in China had been called into question since the mid-1980s. But it would take almost another decade before material changes to be made to improve the financial health of the pension system. And as can be seen in the following discussion, such changes were again incrementally incurred to modify the pension system one small step at a time.

The 3rd Plenary Session of the 14th the Central Committee of the Communist Party of China passed the "Decision of the Central Committee of the Communist Party of China on Some Issues concerning the Establishment of the Socialist Market Economy" on 14 November 1993. Pension reform was singled out as an important avenue for alleviating the financial burdens on enterprises and improving their profitability and competitiveness (CPC Central Committee 1993). With such endorsement by the central government on the direction of pension reform, as a countermeasure for the impending financial stress on the pension system in China caused by populating ageing, various options to reduce financial burdens on the pension system were mooted. Benefits level was an obvious target and different provinces and line ministries would come up with proposals on this aspect of pension reform during this period.

A drop in the pension benefits level as measured by the replacement ratio was first implemented in an economically significant area of the country, as the government of Guangdong issued the "Provisional Regulations on the Old-Age Social Insurance for Employees in Guangdong" on 7 June 1993. According to the policy document, individual accounts for employees in Guangdong were set up. As a start, employees were required to pay 2% of the salaries into their individual accounts, and the personal contribution rate would be adjusted higher later with salary increase. The pension benefits would consist of three components. The accumulation of funds in the individual accounts would fund the first component. The second component would equal to 30% of the average salary of employees in the previous year in the county or city where the retirees were based. The third component would be contingent on whether a retiree had a minimum employment history of ten years. For every year of additional employment beyond ten years, an additional amount of benefits equal to a percentage (from 1% to 1.2%) of the indexed monthly salary averaged across the years of employment would be paid to the retiree (People's Government of Guangdong 1993).

Based on the experience of reform experiments carried out in two cities in Jiangxi and Liaoning, the Ministry of Labour issued the "Notification by the Ministry of Labour on the Work for the Pilot Reform of the Rules of Benefits Determination for the Basic Pension" on 19 October 1993. A formula for determining pension benefits was proposed by the policy document, according to which the pension benefits would consist of two components. The first component would equal to a percentage (from 15% to 25%) of the average monthly salary in the last year prior to retirement. The second component would equal to 1% of the indexed monthly salary averaged across the years of employment multiplied by the number of years of employment (Ministry of Labour 1993).

Learning from the experience of countries such as Singapore and Chile where fully funded pension systems were established, the governments in more advanced urban areas of China were inclined to build up the individualised accounts as a significant source of the pension benefits. For example, the government of Shanghai issued the "Measures for the Old-Age Insurance for Urban Employees in Shanghai" on 27 April 1994. The policy document prescribed to set up individual accounts for urban employees besides the pooled accounts. An employee was required to pay 16% of the salary into the individual account. A substantial component of the pension benefits for a retiree would come from the accumulation of funds in the personal account belonging to the retiree, as a monthly payment equal to the total value of funds in the individual account divided by 120 would be made to the retiree as part of the pension benefits (People's Government of Shanghai 1994).

With the three quite different methods for calculating the pension benefits as proposed by two provincial level governments and one ministry, the central government did not make an immediate, clear-cut decision on which way the pension benefits would be determined nationwide. The State Council issued the "Notification on Deepening the Reform of the Old-Age Insurance System for Employees in Enterprises" on 1 March 1995. Interestingly, two implementation methods were contained in the policy document. They differed substantially in how to calculate the benefits. The first implementation method prescribed that the benefits would be mostly determined by the individual accounts, into which 16% of the salaries would be channelled as pension contributions. The second implementation method insisted on a major role for the pooled account to determine the benefits, with a much smaller individual account for each employee as only 3% to 5% of the salaries would be paid into it (State Council 1995). It was worth noting that the first implementation method was inspired by the reform experiment carried out in Shanghai and supported by the National Commission for Restructuring the Economic System, whereas the second implementation method was recommended by the Ministry of Labour (Xu and Wan 2008). It renders the policy document a piece of historical evidence of ministry rivalry in China. One would have to wait for another two years before seeing how such rivalry in this particular case was resolved by the central government.

2.1.4 More patchworks to the pension system in a phase of fiscal recovery (1997–2005)

A fiscal recovery started to take place since 1997, when the share of government budget revenues in GDP began to increase gradually in China. But the ensuing improved fiscal space did not catalyse a paradigm shift in the way the fiscal reform was conducted, as it still proceeded in a piecemeal, incremental, fragmentary and inadequate manner with accentuated bias on the urban sector (Wong 2007a; Shue and Wong 2007). The pension reform was no exception in this period to the general lack of overall vision and strategy as observed in the fiscal reform. Despite the numerous reform initiatives aimed to improve the existing pension system in one aspect or another as documented in the discussion below, the pension system remained highly fragmented and decentralised. Several thousand separate social insurance funds for the pension system were managed by local governments, often at city or county level, and these local governments were expected to cover the deficits arising from their pension pools, resulting in substantial regional disparities in benefits level across counties even within the same province, since the strength of local public finance and the distribution of pension contributors and retirees could vary widely across the counties (Hussain 2007).

Such deep-seated situation of fragmentation and decentralisation in the landscape of public finance in China is not conducive to achieving an efficient and equitable delivery of the vital public services such as healthcare, education and social security (Wong 2007a). In the case of the pension system, the fragmented and decentralised administration also means sub-optimally low level of risk pooling, which runs against what would be prescribed by the law of large numbers, the most important underlying principle to ensure the financial viability of any insurance plans including pension schemes. The pension reform during this phase of fiscal recovery involved more patchworks to modify the existing system by altering some parameters of the pension system. Regrettably, it lacked the resolve to take a comprehensive approach to defragment the system by designing and implementing a centralised system at the national level and therefore failed to address the real elephant in the room.

The State Council issued the "Decision of the State Council on the Establishment of a Unified Basic Old-Age Insurance System for Employees of Enterprises" on 16 July 1997. A middle ground was reached for the size of the individual account and its importance as a source of pension benefits, as the policy document prescribed that 11% of the salary of an employee should be paid into it. The monthly pension benefits would consist of two components. The first component would equal to 20% of the average monthly wage in the province or prefecture in the previous year. The second component would equal to the total value of funds in the individual account divided by 120. An employee who made pension contributions for at least 15 years would be entitled for pension benefits. For those employees who made pension contributions for less than 15 years, they would not receive pension benefits but be paid a lump sum of money equivalent to the total value of funds in their individual accounts. Enterprises were required to contribute up to 20% of total salaries of their employees, and the provincial governments were given the right to set the exact rate imposed on enterprises. Starting from 1997, employees were required to pay at least 4% of their salaries as pension contributions, and the rate for employees would increase by one percentage point every two years thereafter to eventually reach 8%. All the contributions made by employees would be channelled into their individual accounts, with the rest of the 11% of salaries required for the individual accounts coming from the contributions made by enterprises. The remaining portion of the contributions made by enterprises would be channelled into a pooled account to fund the PAYG component of the pension benefits (State Council 1997). It would prove to be a highly significant policy document in the evolution of pension policy in China, not only because it provided a solution to the above-mentioned competition of ideas between ministries on pension benefits calculation, but more importantly due to the fact that it laid out the framework for a pension system for enterprises employees universally applicable throughout the country.

The implementation of the mixed funding mechanism for the pension system as established by the 1997 reform demands a solution for repaying the implicit pension debt (IPD) incurred when a pension system previously based purely on PAYG financing is transitioning towards one that becomes at least partially funded. The need to settle the IPD, or the transition costs, is generated because at least some of the contributions made by enterprises or individuals must be channelled into individual accounts rather than being used to pay the pension benefits to the currently retired as under PAYG. The IPD for China to transition towards a fully funded pension system is estimated in a report by the World Bank to be 46% to 69% of GDP in 1994 (World Bank 1997). The same World Bank report claims that China had more advantages in its position to resolve the problem of the IPD, but to redeem such potential capacity requires a unification of the fragmented pension system (World Bank 1997).

The unification, of course, did not happen then; it remains a tantalising possibility even to this day. The individual accounts have remained largely empty as the funds are used up for paying the pension benefits, partly due to the financial pressures on the local governments and the lack of determination by the central government to solve the problem of the IPD. The de facto financing mechanism for China's pension system is therefore still PAYG, rather than the mixed financing with a sizable component of funded individual accounts as stated by the policy documents since the 1997 reform (Sin and Yu 2005; Sun and Jiang 2016). Again, the fragmentation and decentralisation in the management of the pension system seemed to have complicated the process for achieving the national policy objective, which in this case involves building a mixed funding mechanism for the pension system.

The development of the pension system in subsequent years would proceed based on the same framework set out in the 1997 reform with a series of parametric adjustments. The State Council issued the "Notification of the State Council on Issuing the Pilot Programme for the Improvement of the Urban Social Security System" on 25 December 2000, modifying the rules over pension contributions such that enterprises would pay about 20% of total salaries of their employees, all into the pooled account, while employees would pay 8% of their salaries, all into their individual accounts (State Council 2000). To promote the participation of people working in the informal sector in the urban areas with unstable employment, the Ministry of Labour and Social Security provided supplementary policies specifically catering for this group of people by issuing the "Notification on Some Issues concerning Improving the Policy of Basic Old-Age Insurance for Urban Workers" on 22 December 2001 (Ministry of Labour and Social Security 2001).

The framework of the pension system established by the 1997 reform was reaffirmed by the top leadership in China when the 3rd Plenary Session of the 16th Central Committee of the Communist Party of China approved the "Decision of the Central Committee of the Communist Party of China on Some Issues concerning the Improvement of the Socialist Market Economy" on 14 October

2003. The policy document made clear that the pension system should expand to include more employees and people in unstable employment in urban areas, and that the individual accounts should be made truly funded with accumulation of assets rather than being used up to pay pension benefits to the currently retired. It prescribed that the management of the pooling of pension funds at prefectural level should be improved in the short term and the pooling of pension funds should then be gradually implemented at provincial level. It also set out the goal of eventually achieving a nationally unified pension system when the right conditions are met (CPC Central Committee 2003).

The pension reform efforts in this era were institutionalised when the State Council issued the "Decision of the State Council on Improving the Basic Old-Age Insurance System for Employees of Enterprises" on 3 December 2005. According to the policy document, all the employees, the self-employed and the people in unstable employment in urban areas should be covered by the pension system. Starting from 1 January 2006, the scale of the individual accounts was officially downsized from 11% to 8% of salaries, all funded by contributions made by the individuals participating in the pension system; the situation of having empty individual accounts should be averted by gradually making them truly funded with assets (State Council 2005).

2.1.5 Pension reform in the context of building a "harmonious society" (2006 to the late 2010s)

The pension reform in China through the 1980s and 1990s until the mid-2000s had proceeded by focusing on adjusting the pension system for urban employees. A common goal of these incremental reform measures as discussed above was to make the system financially more affordable by changing the benefits levels and contribution rates, thus alleviating the burdens on enterprises and facilitating the reform of the state-owned enterprises. These adjustments to the pension system were largely congruent with an attitude adopted by the central government in this period to favour economic efficiency over social fairness.[2] Coupled with the fragmented and decentralised administration of the pension system, such piecemeal adjustments were unable to make the pension system more progressive to help reverse the trend of rising inequality that characterises the social impact of the economic reform in China during this period. The retirees in poorer regions with more rural elements in their local economy received lower level of pension benefits in general, in addition to having to struggle to satisfy their basic needs in times of pension arrears, which tended to happen more often in those poorer regions due to their weaker finances (Wong 2007a; Fock and Wong 2008). Thanks to the lax fiscal management regime imposed on the local governments, a salient feature of the fragmented fiscal system in China, the local pension funds were in some cases even illegally used by the local governments for speculation in shares of companies traded on the stock market and other types of financial securities (Wong 2007b). It was estimated that over 1.5 million retirees, or close to 5% of all retirees across the country, had their pension benefits in arrears or received reduced benefits in 1995 (Zhu 1998). The number of "mass incidents" that could take the form of various kinds of protests increased dramatically from 1993 to 2003, with the problem of pension arrears identified as a major cause for such increase in social unrest (Tanner 2005).

To react to the mounting pressure from the rising social unrest, the central government decided to shift its attitude on the relative importance of economic efficiency and social fairness shortly after the change of top leadership in 2003 with Hu Jintao and Wen Jiabao appointed as the President and the Premier of China, respectively. The 4th Plenary Session of the 16th Central Committee of the Communist Party of China passed the "Decision of the Central Committee of the Communist Party of China on Strengthening the Party's Governance Capability" on 19 September 2004, in which the goal of building a "harmonious society" was first set (CPC Central Committee 2004). In 2005, the 5th Plenary Session of the 16th Central Committee of the Communist Party of China further espoused the importance of building the "harmonious society" and asked for "paying more attention to social fairness in order to let all the people share the results of reform and development" (CPC Central Committee 2005). Recognising the need to restructure the economy to make its growth more sustainable and more driven by domestic consumption, the central government injected substantially more resources, made available by continued economic growth and increase in government revenues in this period, into implementing some impressive and ambitious reform programmes aimed at improving the social safety net and public services provision (Wong 2010).

Under the rubric of "harmonious society", a series of pension reforms was carried out with more fiscal subsidies to improve the pension benefits level and to extend the coverage of the pension system to include rural residents and urban residents without formal employment, as documented below in this subsection. Breakthrough would be made on reforming the pension system for civil servants and public institutions employees. The current pension system consisting of three pension schemes, respectively, covering urban enterprises employees, urban and rural residents, and civil servants and public institutions employees would take shape in this era. However, little progress would be made in this period to extensively revamp the fragmented and decentralised pension system, which still epitomises the dysfunctional Chinese intergovernmental fiscal system that is incompatible for the proper financing of the public services (Wong 2010).

Following the framework of the pension system established by the 1997 reform, the pension scheme for urban enterprises employees, namely the Basic Old-Age Insurance for Urban Employees, continued to develop during this period with some incremental adjustments. For example, to promote labour mobility across provinces and to ensure the transferability of pension contribution history of employees and their pension entitlements, the "Interim Measures for the Transfer of Basic Old-Age Insurance for Employees in Urban Enterprises" was jointly drafted by the Ministry of Human Resources and Social Security and the Ministry of Finance, and passed by the State Council on 28 December 2009 (Ministry of Human Resources and Social Security and Ministry of Finance 2009). Due to the more abundant fiscal subsidies made into the pension system, the benefits level for retirees from urban enterprises also saw consistent improvement, increasing at an average annual growth rate of 12.8% from 2005 to 2016.[3]

The status of the pension scheme for urban employees was further cemented when the "Social Insurance Law of the PRC" was ratified on 28 October 2010 and in force from 1 July 2011, making the requirement for pension contributions by the enterprises and employees legally binding and providing legal protection over pension entitlements of retirees.[4] However, the details concerning the other two pension schemes, respectively, covering urban and rural residents and civil servants and public institutions employees were not given in the 2010 legislation on social insurances, which indicates the presence of an institutional chasm separating the urban enterprises employees with the latter two groups of people (National People's Congress 2010a).

Pension in the form of social insurance was not accessible to most rural residents until very recently in China.[5] On 1 September 2009, the "Guidance of the State Council on the Pilot Reform of the New Rural Old-Age Social Insurance" was released, signifying the establishment of a nationwide pension scheme specifically designed for the rural residents (State Council 2009). Another pension scheme catering for the urban residents, defined here as those with urban registration of household (or urban hukou) but unable or not suitable to participate in the Basic Old-Age Insurance for Urban Employees, was designed and tested with the State Council releasing the "Guidance of the State Council on the Pilot Reform of Old-Age Social Insurance for Urban Residents" on 7 June 2011 (State Council 2011). But these two pension plans only existed briefly as separate schemes of the pension system. By the decision of the State Council, the two schemes were merged to form the Basic Old-Age Insurance for Urban and Rural Residents with the release of the "Opinion of the State Council on the Establishment of a Unified Basic Old-Age Insurance System for Urban and Rural Residents" on 21 February 2014. The scheme would now cover the rural residents as well as the urban residents who were not participating in the scheme for urban enterprises employees (State Council 2014b). As this group of people are usually not formally employed and tend to receive much lower and unstable income as compared to those with formal employment in enterprises, the governments or the public institutions, the financing of the pension scheme was heavily subsidised by central and local governments. However, it only provided very limited pension benefits for the urban and rural residents who retired under this scheme.

In contrast to the substantial progress made on reforming the pension system for urban enterprises employees, the pension plan for civil servants and public institutions employees had remained essentially unchanged for a long time since the reform and opening up started in the late 1970s. Even after three decades of the economic reform in China, the pension plan for this group of people still retained all the hallmarks of the pension system in the early years of the PRC when the country was run as a command economy. Civil servants or public institutions employees and the work units employing them were not required to pay any contributions to help finance the benefits for the retirees under this pension plan. Instead, the pension benefits were entirely paid through public coffers of the central and local governments. The benefits level was also much higher for this group of retirees as compared to their counterparts retiring from urban enterprises, both in absolute terms and in replacement ratio.

Such disparity in the rules over pension contributions and benefits level between urban enterprises employees and this group of people working in the public sector has been perceived as a major social injustice in China. The situation finally started to change with the release of the "State Council Measures for the Pilot Programme on the Reform of the Old-Age Insurance System for Public Institutions Employees" on 14 March 2008. The pilot reform was carried out in Shanxi, Shanghai, Zhejiang, Guangdong and Chongqing to require the public institutions employees and their work units to make pension contributions and to link the pension benefits with contribution history of the employees (State Council 2008). But such experiment was limited in scope and scale. It took almost another seven years for the central government to make a decisive move to reform the pension scheme for civil servants and public institutions employees, when the "Decision of the State Council on the Reform of the Old-Age Insurance System for Civil Servants and Public Institutions Employees" was promulgated on 3 January 2015.

The reform was aimed to establish a pension scheme for civil servants and public institutions employees and to make the pension scheme financially independent from the government agencies and public institutions. The same rules over pension contribution rates and benefits level as per those prescribed by the Social Insurance Law for urban enterprises employees would be applied to the pension scheme for civil servants and public institutions employees after the reform, thereby removing the discrepancy over the pension rules concerning contributions and benefits between the two pension schemes. However, the financing for the pension scheme for the public sector employees remained fragmented and was kept separate from that for the pension scheme for the employees of urban enterprises (State Council 2015a).

2.2 Current system and its schemes

After seven decades of development since its inception and through quite a few twists and turns as documented above, the pension system in China has evolved into the present form that consists of three schemes, namely the Basic Old-Age Insurance for Urban Employees, or the Urban Employees' Scheme (UES) for short, the Basic Old-Age Insurance for Urban and Rural Residents, or the Basic Residents' Scheme (BRS), and the Old-Age Insurance for Civil Servants and Public Institutions Employees, or the Public Employees' Scheme (PES). As their names imply, the three schemes are designed to cater for three different groups of people distinguished mainly by the type of their occupations. The three schemes differ substantially in terms of eligibility, number of participants, contribution rates, benefits level and funding sources. Table 2.1 presents a summary of some key features and statistics of the three schemes.

According to the rules set out by the Social Insurance Law, employees of urban enterprises should join the UES; those self-employed urban residents may join the scheme by making contributions into the pooled pension funds and individual accounts (National People's Congress 2010a). Those urban and rural residents above

Table 2.1 Key features and statistics of the three schemes of the pension system in China, as of the end of 2019

Type of pension scheme	UES	BRS	PES
Eligibility for enrolment	Employees of urban enterprises; those self-employed may join as individual participants	Urban and rural residents above 16 years old and without formal employment	Civil servants and employees in public institutions
Enrolment (million)	434.9	532.7	About 48.0
Compulsory enrolment	Yes	No	Yes
Contribution rates for employers and employees (% of salaries)	16% from employers and 8% from employees; 20% from the self-employed	Fixed annual payments made into individual accounts with government subsidies	16% from employers; 8% from employees
Minimum duration of contribution for pension eligibility	15 years	15 years	15 years
Replacement ratio	59.2% for an employee retiring with 35 years of employment	About 15% to 30%	About 80% to 90%
Average monthly benefits (yuan)	3,333	162	5,941
Source of funding for scheme[13]	81.5% from contributions from employers and employees; 18.5% from subsidies by central and local governments	24.5% from contributions from participants; 75.5% from subsidies by central and local governments	Central and local governments

Sources: 2017 Statistical Bulletin on the Development of Human Resources and Social Security (Ministry of Human Resources and Social Security 2018, 2020), China Labour Statistical Yearbook 2018 (National Bureau of Statistics 2018a), and relevant State Council policy documents (State Council 2005, 2014b, 2015a).

16 years of age and without formal employment are eligible for participating in the BRS (State Council 2014b). The employees working in the governments and its agencies as well as the public institutions should join the PES (State Council 2015a).

The UES has achieved an enrolment of 434.9 million by the end of 2019, which includes 311.8 million working age participants and 123.1 million retirees. Despite its enrolment being voluntary unlike that of the other two schemes, with 532.7 million participants in 2019, the BRS provides pension protection for more people than the other two schemes put together (Ministry of Human Resources and Social Security 2020). The smallest scheme in terms of the number of participants, the PES is estimated to cover around 48 million public sector employees in total.[6]

The same contribution rates for employers (16% of the payroll made into the pooled pension funds) and employees (8% of their salaries made into individual accounts) are applicable for both the UES and the PES, with the exception for the self-employed participants of the former scheme, who need to pay 20% of a reference salary set by the local authority in charge of the pension system with 8% going into the individual accounts and the other 12% going into the pooled pension funds (State Council 2005, 2015a).[7] The participants of the BRS need to make fixed annual payments into their individual accounts supplemented by subsidies from the central and local governments. The required fixed annual payments are low as compared to the other two schemes and set in multiples of 100 yuan ranging from 100 yuan to 1,000 yuan with two higher permissible levels at 1,500 yuan and 2,000 yuan; participants of the scheme have the right to choose the level of contribution and the local governments have the discretion to add additional contribution levels besides the 12 levels (State Council 2014b).

The pension benefits level varies significantly among the three schemes. The UES is designed to provide a replacement ratio of 59.2% for an employee retiring with 35 years of contribution history (State Council 2005), and the average monthly benefit of the scheme in 2019 was 3,333 yuan (Ministry of Human Resources and Social Security 2020). The BRS only provides limited pension benefits for its retirees with the average amount of monthly benefits at 162 yuan in 2019 (Ministry of Human Resources and Social Security 2020); in the case of the rural residents, the replacement ratio, defined as the pension benefits as a proportion of the net income of the rural residents, ranges between 15% and 30% (Feng 2016). The pension benefits level afforded by the PES is the most generous. The estimated amount of the average monthly benefits for the scheme in 2019 is 5,941 yuan,[8] with its replacement ratio typically ranging between 80% and 90% (ZhigangYuan et al. 2016). Such statistics indicate that the retired civil servants and public institutions employees typically get almost twice as much of pension benefits as the enterprises retirees and that there is a difference in magnitude in the benefits level between those retired under the scheme for urban and rural residents and those retired under the other two schemes, as the average amount of benefits for urban and rural residents is about 4.9% and 2.7% of that of the other two schemes, respectively.

The funding structure of the three schemes also differs considerably. The majority of the funding for the UES comes from contributions made by employers and employees, with government subsidies being an important supplement to the total revenues for the scheme; employers and employees contributed 81.5% of the total revenues of the scheme with subsidies from central and local governments making up 18.5% of the total revenues in 2017 (Ministry of Human Resources and Social Security 2018). The reverse is true for the BRS, with government subsidies constituting most of the revenues for the scheme; in 2017, 75.5% of the total revenues of the scheme came from subsidies by central and local governments with the participants of the scheme contributing 24.5% of the total revenues of the scheme (Ministry of Human Resources and Social Security 2018).With the state being the ultimate employer of the participants of the PES, the payroll costs of this group of employees are borne by public finance. Therefore, the source of funding for the

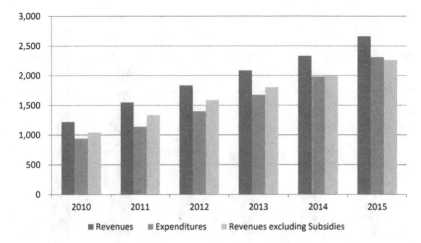

Figure 2.2 Revenues and expenditures (in billion yuan) for the UES, 2010–2015.

Sources: Annual Report on China's Social Insurance Development in 2015 (Ministry of Human Resources and Social Security 2016b).

PES is provided by central and local governments, although the employees and work units covered by the scheme are now required to contribute at the same rates as per their counterparts in enterprises after the 2015 reform (Zhigang Yuan et al. 2016).

With the rapidly ageing population, the financial sustainability of the Chinse pension system has attracted much public concerns and scholarly interest in recent years. In the short term, the financing for both the UES and the BRS still remains quite robust. At the end of 2017, the former scheme had accumulated a surplus totalling 4.39 trillion yuan, enough to cover for 13.8 months of the pension benefits under the scheme at that time, with the latter scheme also having accumulated a sizable surplus at 632 billion yuan (Ministry of Human Resources and Social Security 2018). But the medium- to long-term prospect of the financial health of the system is less than reassuring. As shown in Figures 2.2 and 2.3, after removing government subsidies, the annual expenditures started to exceed the annual revenues for the UES in 2015, and such deficit emerged as early as in 2011 for the BRS. In other words, without further reform, it has been already the case since 2015 that the total annual contributions received by the pension system are unable to cover the total benefits due in the same year.

The present form of the Chinese pension system consisting of the three schemes is the result of the numerous reform efforts and initiatives taken by the central and local governments as well as the line ministries throughout the decades. The system now delivers pension protection for the vast majority of the people in the country regardless of their occupation and no matter whether they are urban or rural residents, albeit at very different benefits levels. The merge of the urban and rural residents' pension schemes and the nominal removal of discrepancy in pension contribution rules between the pension scheme for enterprises employees and that for civil servants and public institutions employees represent some bold moves towards

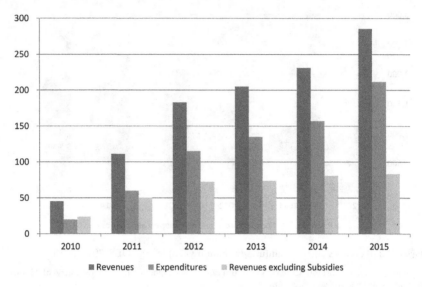

Figure 2.3 Revenues and expenditures (in billion yuan) for the BRS, 2010–2015.

Sources: Annual Report on China's Social Insurance Development in 2015 (Ministry of Human Resources and Social Security 2016b).

unifying the pension schemes in China. Viewed from the historical context, huge achievement has been made in its development. However, these pension reforms have been carried out in a piecemeal, reactive manner and the system is still far from an optimal one. The root cause for its many drawbacks can be traced to its highly frag-mented and decentralised administration, which has been a persistent attribute of not only the pension system but also the fiscal system in China in general.

Notes

1 In fact, it would take more than five decades since the establishment of the urban pen-sion system before a substantial proportion of the rural population would be entitled to pension in any form or shape, as even by 2003, only about 7% of the rural population in China were covered by some kinds of pension schemes (State Council Information Office 2004).
2 The attitude on the relative importance of efficiency and fairness with implications for guiding policies on income distribution was endorsed by the 3rd Plenary Session of the 14th the Central Committee of the Communist Party of China and documented in the "Decision of the Central Committee of the Communist Party of China on Some Issues concerning the Establishment of the Socialist Market Economy" (CPC Central Committee 1993).
3 Fiscal subsidies allocated to help finance the Basic Old-Age Insurance for Urban Employees increased substantially after 2005. The amount of such subsides was 40.8 bil-lion yuan in 2002, and reached 651.1 billion yuan by 2016, registering an average annual growth of 23.7% during the six years from 2010 to 2016 (Caijing 2017).
4 A draft of the Social Insurance Law was passed in principle by the State Council on 28 November 2007 (State Council 2007). Besides the old-age social insurance (i.e. pen-sion), the law also provides for the rules and regulations for four other types of social

insurances including medical insurance, work injury insurance, unemployment insurance and childbirth insurance, which collectively constitute an important part of the social safety net in China.

5 In the pre-reform era, pension provision in rural areas of China was rare, sporadic and locally based, often restricted to residents living in the same villages with different rules and benefits levels in different villages if pension was available there at all. Since the 1980s, some efforts to set up a national rural pension scheme were made, but no substantial progress was achieved to extend the pension protection to cover any significant portion of the rural population. By the end of 2003, more than 90% of the rural population in China were not participating in any pension scheme (State Council Information Office 2004).

6 Zhu Lijia, a scholar from the Chinese Academy of Governance, estimated the number of people salaried by public finance to be around 50 million (People's Daily 2016). Currently there are around 2 million armed services personnel in China (Jian Xiong 2016), who are also salaried by public finance. Excluding the armed services personnel from Zhu's estimation gives the estimated number of public sector employees at around 48 million.

7 The contribution rates for the UES were first set in the 1997 reform of the pension system. The contribution rate for employees of 8% of salaries has remained the same and is standard across the country, whereas the contribution rate for employers has some regional variations. According to the "Decision of the State Council on the Establishment of a Unified Basic Old-Age Insurance System for Employees of Enterprises", provinces were allowed to levy up to 20% of payrolls on employers, and local governments were given the right to set the exact rate imposed on employers; provinces were allowed to set the rate higher than 20% if they needed to do so to pay for the pension benefits due to higher proportion of retirees, but they had to propose the plan and seek approval from the Ministry of Labour and the Ministry of Finance (State Council 1997). The central government made some adjustments in recent years to allow the provinces to lower the contribution rate for employers, first from 20% to 19% for provinces with enough pension surpluses to cover more than 9 months of pension benefits in 2016, and then to 16% for all provinces in 2019 (Ministry of Human Resources and Social Security and Ministry of Finance 2016b, 2018; State Council 2019). When the economy was hit hard during the outbreak of COVID-19 in early 2020, the central government decided to implement a significant social insurance fee relief programme starting from February 2020 (Ministry of Human Resources and Social Security, Ministry of Finance, and State Taxation Administration 2020). Small- to medium-sized firms and microbusinesses were temporarily exempted from making pension contributions, while large firms were allowed to pay at half of the normal rate. In 2020, the programme resulted in a total of 1.33 trillion yuan of fee relief, equivalent to about a quarter of the total revenues for the pension system in China in 2019 (State Council Information Office 2021).

8 According to Wang et al., data about income of civil servants has been confidential since 2006 (Wang, Béland, and Zhang 2014a). The average amount of monthly pension benefits for civil servants and public institutions employees is estimated to be 5,941 yuan in 2019, based on the estimation by Wang et al. that the average annual pension income for civil servants in China in 2011 was 40,060.25 yuan (Wang, Béland, and Zhang 2014a) and the assumption that the pension benefits for civil servants and public institutions employees grew at 10% annually from 2011 to 2015 and at 5% afterwards until 2019. The pension benefits for retirees under the UES grew at an average annual rate of 10.3% from 2011 to 2015 (Ministry of Human Resources and Social Security 2016b), and at about 6.5% in 2016, 5.5% in 2017 and 5% in 2018 and 2019 (National Business Daily 2021). So, the assumption of 10% annual growth from 2011 to 2015 and 5% afterwards until 2019 was a plausible one.

3 Population ageing and pension sustainability in China

This chapter studies the impact of population ageing on pension sustainability in China based on the notion of de facto bankruptcy, a more nuanced conception of the financial sustainability of the pension system. Treating the pension system as a nationally unified scheme, the chapter assesses whether the long-term financial sustainability of the system can be ensured through retirement age reform under different demographic scenarios. Financing methods such as fully funded and PAYG plans are also tested to find out which financing method generates the optimal welfare effect in each retirement age reform option under each demographic scenario. The possibility of the fertility cliff is explored with a wider range of fertility scenarios. The first section reviews the background of the problem presented by the demographic challenge. The second section describes the methodology and research design. The third section presents and discusses the results of the quantitative simulation and scenario analysis. The last section highlights the key findings of the analysis presented in the chapter.

3.1 Background of the demographic challenge

China had benefited from a period of demographic dividends during the first three decades of its reform towards a market economy since the late 1970s. The total dependency ratio of the country kept decreasing from 1970 until 2010 (United Nations 2019b). Now, it appears that the period of demographic dividends has ended. China is facing the severe issue of population ageing. According to the projections by the Population Division of the United Nations, the ratio of young and elderly dependents in the total population in China will increase to around 50% by the middle of the century, when the number of working-age adults will be roughly equal to the number of dependents in the country (United Nations 2019a).[1]

Two forces are responsible for creating the demographic dividends China enjoyed and the ensuing demographic pressure it faces from population ageing occurring at an unprecedented speed and scale. First, life expectancy in China has greatly improved in the past few decades, converging towards that of the developed countries. There used to be a gap of 20 years in life expectancy between China and the average of that in OECD countries in 1960; in 2014, this gap had shrunk to be less than five years, as can be seen in Figure 3.1. China presently has a life expectancy of around 77 years (World Bank 2020b). Second, the fertility rate in China has

DOI: 10.4324/9781003182696-3

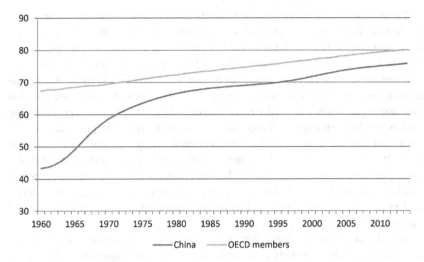

Figure 3.1 Life expectancy: China vs. OECD countries, 1960–2014.

Sources: The World Bank (2016).

experienced an even more dramatic change. Chinese women used to have more than six children on average in the 1960s. But that number had decreased drastically by the early 1970s, before the implementation of the famous one-child policy. In the late 1970s, when the one-child policy was adopted, there was a further drop in fertility level. Since 1992, the total fertility rate has been below the replacement level in China (World Bank 2021). It is worth noting that similar downward trends in fertility have been observed in many neighbouring countries in East Asia and that such declines can be ongoing for quite a long period such as half a century or even longer. These countries did not implement the one-child policy; yet their fertility rates all dropped in a similar way when they became more prosperous economically.

With the fertility rate now reaching a very low level, China may face the problem of a low fertility trap, where the prevailing low fertility levels engender further declines in fertility.[2] With such a threat of the low fertility trap, the Chinese government started to relax controls on childbirth in 2013 by allowing couples to have a second child if either parent was a single child (CPC Central Committee 2013). However, few chose to do so; it was estimated that only 7% of eligible couples applied for the permit to have a second child (CCTV America 2015). On 29 October 2015, the Central Committee of the Communist Party of China abolished the one-child policy, which had lasted for almost 40 years (CPC Central Committee 2015a). But some birth control still remains in China, with a "two-child" policy being in force now (National People's Congress 2015). It is widely believed that China's birth rate will not increase substantially with the end of one-child policy (Schiermeier 2015; Elliott 2015). Indeed, China may even experience a further fertility decline in the foreseeable future before seeing any upward turn in birth rates. Such a scenario, if it turns out to be the case, is actually following the "norm" of the demographic trends of other East Asian economies including Japan, Korea and even Thailand and Vietnam to a lesser extent.[3]

There is no apparent reason to believe that China, once entering the so-called low fertility trap (CCTV America 2015), can escape easily in a timely manner once the government relaxes its birth control. Two years after the one-child policy was abolished, the birth rate in China reached 10.94 per 1,000 people in 2018, the lowest since 1949 (National Bureau of Statistics 2019a). This is particularly concerning since the effect of pronatalist measures has been shown to be not that encouraging, as borne out in the recent history of East Asian countries. To escape the low fertility trap requires substantial social changes, which require considerable amount of time and often lag behind the economic changes (Ioannides and Nielsen 2007). It may take a few decades, if not an entire century, to see the reversal in the change in fertility rate.[4] If China is indeed trapped in the low fertility trap, there is a high likelihood that it will stay in the trap for a long period of time, as no country to date has subsequently increased its fertility rate back to the replacement level after falling into the low fertility trap.

The threat of the low fertility trap prompts us to consider the existence of a fertility cliff in China, namely a threshold of fertility rate below which all pension schemes would be in de facto bankruptcy. Theoretically, any pension scheme can be made sustainable, by adjusting either the contribution rate, the benefits level, the retirement age, or a combination of those. But a pension system should be considered to be in de facto bankruptcy if it requires unrealistically high contribution rate, or miserably low benefits, or exorbitantly high retirement age. Consequently, the sustainability of the pension system in China can be judged in its ability to prevent de facto bankruptcy, which not only depends on the internal system parameters but also is contingent on whether such a fertility cliff exists and, if so, where China is now in relation to the fertility cliff.

This chapter focuses on addressing the first research question as stated in Chapter 1 by studying the implications of demographic changes for the financial sustainability of the Chinese pension system and assessing the potential for delaying the mandatory retirement age as a policy response. Specifically, this chapter attempts to seek answers to the following research questions. First, what impact will the fertility trends of China generate on the sustainability of its pension system under the plausible reform options in retirement age? Second, how do different plausible reform options in financing method, retirement age and timing of implementation affect overall welfare outcome across generations while ensuring financial sustainability under different fertility scenarios? And third, is there a fertility cliff under which the pension system will be in de facto bankruptcy with all plausible retirement age reform options?

With the very serious demographic challenges facing China, there is no guarantee that parametric adjustments such as postponing retirement age will ensure the financial health of the pension system by avoiding de facto bankruptcy. To ascertain the boundaries of the requisite demographic conditions within which the long-term sustainability of the pension system could be achieved through adjusting the retirement age, the above-stated research questions are raised.

To answer these questions, three demographic scenarios are constructed to reflect the baseline, optimistic and pessimistic cases of future fertility levels in China and a quantitative approach is adopted for the simulation analysis using an OLG model. The retirement age is used as the main policy variable. Under the different fertility scenarios, quantitative simulation of the retirement age reform is conducted to assess different plausible options for the retirement age and the timing of introducing the postponement in retirement age. The retirement age reform options are compared in terms of the welfare effects to determine the optimal option under each of the fertility scenarios. The long-term financial sustainability is always ensured through parameter adjustment in each round of simulation for the reform, so that the welfare effects generated by these reform options can be compared on an equal footing. In each fertility scenario, under each retirement age reform option, four alternative financing methods for the pension system are considered to compare their welfare effects across generations. They include a fully funded plan and three variations of PAYG-based plans.

3.2 Methodology and research design

As the aim of this chapter is to assess whether the long-term financial sustainability of the pension system in China can be ensured through retirement age reform while abstracting away from the effects of fragmentation by assuming a nationally unified pension system that is amenable to standard macroeconomic modelling tools, the analysis is structured around the three research questions as outlined in Section 3.1. First, the impact of the fertility trends in China on the sustainability of its pension system is studied under the plausible reform options in retirement age. The optimal option for the retirement age reforms in terms of overall welfare outcome across generations is then identified while ensuring financial sustainability under different fertility scenarios. After that, the possibility of a fertility cliff under which the pension system will be in de facto bankruptcy with all plausible retirement age reform options is assessed.

To study the financial implication of the fertility trends on the pension system in China, the demographic scenarios in which to conduct the quantitative analysis of retirement age reform are constructed. Life expectancy of China is very likely to increase further and to converge with that of the developed countries in the coming decades (United Nations 2015). However, there is a much greater degree of uncertainty about the outlook of the total fertility rate in China. Thus, the demographic scenarios are constructed by altering the assumptions about the future total fertility rate in China to assess the sustainability of the pension system. Three fertility scenarios, namely the baseline, optimistic and pessimistic scenarios, are constructed.

The baseline scenario assumes that future fertility profile of China until 2100 follows the medium variant projection by the United Nations (United Nations 2015). After 2100, the total fertility rate starts to increase to reach the replacement level linearly in 50 years to achieve a steady population level in the long run.[5] The optimistic scenario assumes that the total fertility rate increases linearly to the replacement level in one decade and stays stable thereafter. This scenario is intended to represent the upper bound of the plausible range of fertility outcomes, encapsulating the effects of not only the abolishment of one-child policy but also other foreseeable government initiatives for boosting fertility.

The pessimistic scenario posits that the total fertility rate remains significantly below the replacement level for the next few decades and only starts to increase after 2100 to reach the replacement level linearly in 50 years. According to the National Bureau of Statistics, the total fertility rate in China was 1.05 in 2015,[6] which is considered extremely low (National Bureau of Statistics 2016b). A discount on this figure is taken to reach an even more extreme level at 0.9 as the total fertility rate for China from the present time until the end of the century. This is equivalent of assuming that the abolishment of the one-child policy will not increase the total fertility rate in China as much and as quickly as hoped for by the government and that further declines in total fertility rate will occur to keep the rate at sustained low levels, as implied by the low fertility trap hypothesis. Setting the total fertility rate at such extremely low yet not improbable level is necessary to make the pessimistic scenario capture the worst-case scenario for the outlook of fertility levels in China,[7] hence enabling us to probe the likelihood of the fertility cliff.

After constructing the demographic scenarios, the plausible options of reform to be tested by numerical simulation are defined to investigate their viability for ensuring the long-term financial sustainability of the pension system in China. The retirement age is used as the main policy tool to be examined, which is being planned at the moment by the government (BBC 2015). The retirement age in China is about seven years earlier than the OECD average of 64 (OECD 2017). As the life expectancy gap between China and the OECD average is only about four years (World Bank 2019), it indicates that there may be some scope for reforming the retirement age. Under the baseline, optimistic and pessimistic fertility scenarios, two reform options for postponing retirement age are tested. The first is to retire at 60 and the second 65. The current weighted average of the statutory retirement age in China is about 57 (Song et al. 2015). Therefore, the first and second options represent three years and eight years of retirement age postponement, respectively. A three-year postponement in retirement can be considered a lower bound for any meaningful and significant reform of retirement age. As life expectancy in China is around 76 years (World Bank 2019), a retirement age at 65 means 11 years of retirement on average, which is rather short. Therefore, the second option of retirement at 65 represents the upper limit of postponement of retirement age while maintaining political viability.

In terms of the timing of when the postponement in retirement age is implemented, two options are considered. The first is to postpone the retirement age in 2022. According to Yin Weimin, former Minister of Human Resources and Social Security of China, the details of the plan for postponing retirement age were being drafted in 2016, and the implementation of such postponement of the retirement age will start from 2022 at the earliest (BBC 2015). Therefore, the first option represents early implementation. The second is to postpone the retirement age in 2032, which represents late implementation with a waiting period of another decade. China currently is in a phase of rapid population ageing; according to a report by the Chinese Academy of Social Sciences, the share of the population over 65 years old in China will overtake that of Japan and become the highest in the world by 2030 (People's Daily 2010). Therefore, waiting for another ten years before starting to postpone the retirement age in 2032 represents an upper bound for the timing for activating the retirement age reform, if such reform is to be launched at all.

After determining the specification of the demographic scenarios and the details of the options and timing for the retirement age postponement, a modelling method in which to conduct the simulation of the reforms is needed. Due to the quantitative nature of the research questions related to the pension system, a macroeconomic modelling tool that is capable of capturing the effects of changes in the demographic structure of an economy should be adopted.

The OLG models belong to one of the common analytical frameworks for studying the dynamics of economic growth. In the OLG models, people only live for a limited number of periods and they live long enough to coexist with people of other age group for at least one period of their lives. This is unlike other types of neoclassical growth models such as the Ramsey model that are built on the assumption of people living for infinite amount of time. The OLG models are able to capture the impact of the demographic trends on the economy by incorporating their influence on the supply and demand of production factors such as labour. Due to these features of the OLG models, they are the standard tool for analysing the interaction between the demographics, the pension system and the macroeconomic dynamics of a country (Fanti and Gori 2012; Weil 2008; Verbič, Majcen, and Nieuwkoop 2006; Meijdam and Verbon 1997; Pecchenino and Pollard 1997; Breyer 1989).

A recently developed OLG model calibrated to China is adopted to perform an assessment of the likely effects of pension reform options to help identify the optimal reform option based on the welfare criterion adopted by a benevolent social planner, who values the lifetime utility of people across all present and future generations (Song et al. 2015). The adopted OLG model is rich in its features as it is designed to be capable of carrying out analysis on various types of financing methods for pension systems including the PAYG method and the fully funded method, the two primary financing methods commonly used for pension systems in countries across the world. The PAYG system pays the benefits to retirees by using the contributions from current workers and employers. The fully funded system pays the benefits to retirees by using the funds accumulated in personal accounts from previous contributions plus any investment returns. The existing pension system in China adopts a combination of both financing methods according to its policy design (State Council 1997, 2005). In reality, though, the pension benefits have been by and large financed through a PAYG system with the individual accounts being widely empty for the funded component (Zheng 2016).

The model includes a built-in demographic model. The demographic model not only projects the evolution of population dynamics of China based on the detailed data for the distribution of fertility and mortality rates by age and gender groups, but it also takes into account the dynamics of migration from rural to urban areas, which is an important feature of China currently undergoing a massive urbanisation process that substantially affects many aspects of the economy including the pension system. More importantly, differing from the standard OLG models that are built for developed economies in steady state with efficient capital markets, the core part of the model developed by Song et al. realistically captures many salient features and interdependent aspects of the Chinese economy as a developing country, such as the relatively high growth rates for wages and productivity as well as the lack of well-developed capital markets. All these factors and features of the model make it well suited for quantitatively simulating the effects of the different options of pension reforms on public welfare in the context of China.

To improve the accuracy of the forecast of the demographic structure of China, I update the demographic model within the OLG model developed by Song et al. based on the latest census data. The model developed by Song et al. contains a demographic model that is used to make projections of the population dynamics in China, which in turn are fed into the main OLG model as input for the demographic structure of China in the future when performing prospective modelling of the economy under different pension reform options. Their demographic model is built based on the data of fertility rates and mortality rates from the 2000 Census and 2005 One-Percent Population Survey conducted by the National Bureau of Statistics of China (Song et al. 2015). I update the demographic model using the data from the 2000 Census and the 2010 Census, since the latter contains more recent and representative demographic information than the 2005 One-Percent Population Survey. I also derive an updated estimation of the gender-age-specific rural-urban migration rates based on the data from the two censuses, following the same estimation strategy adopted by Song et al.

The updated model is used to carry out simulation analysis on the retirement age reform, with the two options for postponing the retirement age and the two reform timing options, under the three fertility scenarios. Both PAYG and fully funded plans for financing the pension system are examined for each combination of the retirement age options and the reform timing options under each fertility scenario. The long-term financial sustainability is ensured through parameter adjustment on the pension system in each round of simulation for the reforms, so that the comparison of the welfare effects generated by these reform options can be made on an equal footing. The results of the quantitative simulation of retirement age reforms are reported in the next section.

3.3 Results and discussion of quantitative simulation and scenario analysis

The projection of population dynamics in China based on the baseline, optimistic and pessimistic fertility assumptions leads to three demographic scenarios, under which the quantitative simulation of the retirement age reforms is conducted. Figure 3.2 shows the projected population dynamics of China, with Panels A, B and C corresponding to the baseline, optimistic and pessimistic scenarios, respectively. The three solid lines denote the projected total, urban and rural populations, respectively.

In all three scenarios, the total population in China follows the similar trend of first increasing to reach a peak value and then decreasing for the rest of the century. The peak value for the total population and the time at which it is reached vary across the three scenarios. In the baseline scenario, the peak total population of

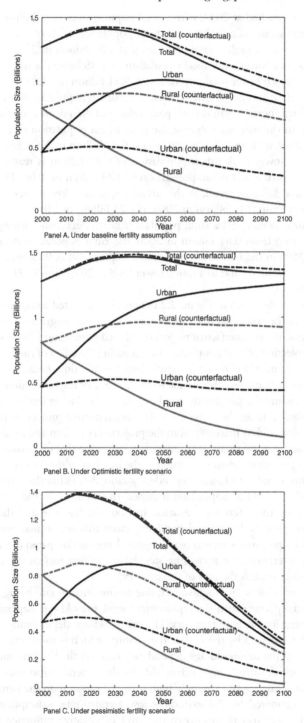

Figure 3.2 Projected population dynamics, 2000 to 2100.

1.41 billion is reached at 2024, with a total population at 0.91 billion in 2100. In the optimistic scenario, a higher peak total population of 1.47 billion is reached later in 2036, with a higher total population at 1.33 billion in 2100. In the pessimistic scenario, a lower peak total population of 1.38 billion is reached earlier in 2015, with a much lower total population at 0.31 billion in 2100.

In the baseline and pessimistic scenarios, the urban population follows the similar trend of first increasing to reach a peak value and then decreasing for the rest of the century. In the baseline scenario, the peak urban population of 1.02 billion is reached in 2049, with an urban population at 0.84 billion in 2100. In the pessimistic scenario, a lower peak urban population of 0.87 billion is reached earlier in 2035, with a much lower urban population at 0.29 billion in 2100. The optimistic scenario has a different trend, as the urban population keeps increasing for the entire century, from 0.48 billion in 2000 to 1.23 billion in 2100.

In all three scenarios, the rural population keeps decreasing during the entire century. Starting from 0.81 billion in 2000, the rural population decreases to 72 million in 2100 in the baseline scenario, to a higher level at 96 million in 2100 in the optimistic scenario, and to a much lower level at 20 million in 2100 in the pessimistic scenario.

The three dotted lines in Figure 3.2 denote the projected total, urban and rural populations in the counterfactual case of zero rural-urban migration in China. In all three scenarios, the counterfactual projection of total population follows similar trend as per the projection of total population incorporating the effect of rural-urban migration, but the counterfactual one is slightly larger at any time because of the higher proportion of the rural population due to the absence of migration from the rural to the urban areas and the fact that the rural population has higher fertility rates than the urban population. In all three scenarios, the counterfactual projection of urban and rural populations differs markedly from the projection of urban and rural populations taking into account the rural-urban migration, indicating that urbanisation is a major driver of the growth of urban population in the first half of the 21st century.

The future trends of old-age dependency ratio, defined as the ratio of population over 60 years old to population between 18 and 59 years old, vary considerably among the three fertility scenarios. In both the baseline and the optimistic scenarios, the projected old-age dependency ratios follow a similar trend, with the values for the optimistic scenario being slightly lower as compared to their baseline scenario counterparts, first increasing steeply in the first half of the century and then stabilising towards the end of the century.

In the pessimistic scenario, however, due to the fertility rate being persistently and substantially lower than the replacement level, the old-age dependency ratios keep increasing for the entire century. The urban old-age dependency ratio reaches 0.69 in 2050 and 1.93 by the end of the century, which is more than three times of the values of the urban old-age dependency ratios in the baseline and optimistic scenarios. Such drastically high urban old-age dependency ratio under the pessimistic scenario implies an extremely high financial pressure on the pension system, which is engendered by the extremely low fertility rate assumption purposely made for this scenario to pin down the limit of negative influence that can be exerted by an adverse fertility trend on pension sustainability.

The current pension system in China with its existing parameters is not financially sustainable for the long term. When simulating the effect of the retirement age reform options, the level of pension benefits is adjusted so that the pension system after the reform achieves long-term sustainability. The financial balancing of the pension system is achieved when the intertemporal budget constraint for the pension system holds with a given level of initial pension fund.[8]

In each round of retirement age reform simulation, the financial balancing of the pension system results in a long-term sustainable replacement rate (LTSR) that can be compared across the different reform options, since the pension system after the retirement age reform is made financially sustainable for the long term in each of the reform options under each demographic scenario. The simulation results of these retirement age reforms under the three demographic scenarios are reported and discussed in the following parts of this section.

3.3.1 Simulation results under the baseline fertility scenario

Under the baseline fertility scenario, the six retirement age reform options yield different values of LTSR, as reported in Table 3.1. The reform options are named a, b, c, d, e and f, with their corresponding combination of retirement age and implementation time shown in Table 3.1. This naming convention is retained in the following parts of the chapter, so that reform option a means the reform implemented in 2022 to financially balance the pension system with the retirement age being 57,[9] reform option b means the reform implemented in 2022 to financially balance the pension system with the retirement age being 60, and so on.

The LTSR increases with the retirement age, as expected. In the baseline fertility scenario, the financial balancing of the pension system carried out in 2022 with a retirement age of 57 results in an LTSR of 29.8%. The LTSR can be increased to 36.7% and 51.5%, if the retirement age is delayed in 2022 to 60 and 65, respectively. With the same retirement age, the ten-year delay in implementation of the financial balancing and retirement age reform decreases the LTSR by 1.7, 1.3 and 0.3 percentage points correspondingly for the retirement age of 57, 60 and 65. So the delay by a decade in launching the reform causes some minor reduction in the LTSR with each retirement age reform option under the baseline fertility scenario.

Table 3.1 LTSR for the pension system under different retirement age reform options under baseline fertility scenario

Reform options	Retirement age	Implementation time	LTSR
a	57	2022	29.8%
b	60	2022	36.7%
c	65	2022	51.5%
d	57	2032	28.1%
e	60	2032	35.4%
f	65	2032	51.2%

Under each retirement age reform option, four plans of financing method reform for the pension system are tested to compare the welfare effects across generations of the four alternative financing plans. The first plan is to keep the PAYG system and to delay the financial balancing to 2050. The second is to keep the existing PAYG system and to delay the financial balancing to 2100. The third is to convert to a fully funded system at the same time when the retirement age reform is implemented in 2022 or 2032. The fourth is to convert to an annually actuarially balanced PAYG system at the same time when the retirement age reform is implemented in 2022 or 2032. In the fourth plan, the benefits level is adjusted annually, so that the total annual revenues equal to total annual expenditures every year.

The welfare effects across generations are summarised by the statistic equivalent variation, as defined in Song et al. (2015). The equivalent variation is the required change in percentage in lifetime consumption of all people across the living generations to make them indifferent between the alternative financing plan (i.e. the first, second, third and fourth plan as described above) and the benchmark plan. The benchmark plan is the default reform plan under each retirement age reform option (i.e. reform options a to f). The equivalent variation is measured in percentage and reported in Table 3.2 for the four alternative financing plans under each retirement age reform option. A higher equivalent variation value implies that the plan generates a more preferred situation for the social planner who values the utilities of all people alive at the time of reform.[10]

For reform options a, b, d and e, under which the retirement age is either 57 or 60 when the reform is implemented in either 2022 or 2032, the best plan among the four alternative financing schemes is to keep the PAYG system and to delay the financial balancing to 2100, as it carries the highest value of equivalent variation. For reform option c and f, under which the retirement age is 65 when the reform is implemented in either 2022 or 2032, the best plan among the four alternative financing schemes is to convert to the annually actuarially balanced PAYG system.

The fully funded financing plan is always dominated by some PAYG-based financing plans in terms of the value of equivalent variation, and it is therefore never the best financing plan for the six retirement age reform options under the baseline fertility scenario. This is not surprising, given that, unlike the PAYG-based

Table 3.2 Comparison of welfare gains (equivalent variation) of pension financing plans under different retirement age reform options under baseline fertility scenario

Retirement age reform options/ retirement age, implementation time	Balancing delayed until 2050	Balancing delayed until 2100	Fully funded	Actuarially balanced annually
a/57, 2022	6.8%	10.5%	0.4%	4.6%
b/60, 2022	5.0%	7.8%	0.2%	7.1%
c/65, 2022	1.5%	2.4%	−0.2%	10.0%
d/57, 2032	3.3%	7.0%	2.2%	1.3%
e/60, 2032	2.6%	5.3%	2.4%	4.6%
f/65, 2032	0.9%	1.8%	2.5%	9.4%

pension systems, a fully funded pension system lacks the ability to redistribute wealth across the generations, which is a valuable feature of the pension system design for the social planner who cares for the welfare of people of all generations. The fact that the wage growth rate would remain higher than the interest rate for years to come in an emerging economy like China also contributes to the PAYG being favoured over the fully funded, since the former generates a return at the rate of aggregate wage growth for the participants of the pension system and the return rate for the latter equals to the interest rate.

3.3.2 Simulation results under the optimistic fertility scenario

Under the optimistic fertility scenario, the six retirement age reform options yield different values of LTSR, as reported in Table 3.3. Similar to the baseline fertility scenario, the LTSR increases with the retirement age, as expected. In the optimistic fertility scenario, the financial balancing of the pension system carried out in 2022 with a retirement age of 57 results in an LTSR of 32.0%. The LTSR can be increased to 38.9% and 54.2%, if the retirement age is delayed to 60 and 65, respectively. With the same retirement age, the ten-year delay in implementation of the financial balancing and retirement age reform decreases the LTSR by 1.0, 0.7 and 0.2 percentage points correspondingly for the retirement age of 57, 60 and 65. So the delay by a decade in launching the reform causes some minor reduction in the LTSR with each retirement age reform option under the optimistic fertility scenario. Compared to the baseline fertility scenario, unsurprisingly, higher values of LTSR are obtained under the optimistic fertility scenario, as the LTSR is on average 2.6 percentage points higher across the six retirement age reform options.

Under each retirement age reform option, the same four alternative financing plans for the pension system are tested as under the baseline fertility scenario to compare the welfare effects across generations of the four financing plans. The welfare effects across generations are summarised by the statistic equivalent variation, which is reported in Table 3.4 for the four alternative financing plans under each retirement age reform option.

For reform options a and d, under which the retirement age is 57 when the reform is implemented in either 2022 or 2032, the best plan among the four

Table 3.3 LTSR for the pension system under different retirement age reform options under optimistic fertility scenario

Reform options	Retirement age	Implementation time	LTSR
a	57	2022	32.0%
b	60	2022	38.9%
c	65	2022	54.2%
d	57	2032	31.0%
e	60	2032	38.2%
f	65	2032	54.0%

Table 3.4 Comparison of welfare gains (equivalent variation) of pension financing plans under different retirement age reform options under optimistic fertility scenario

Retirement age reform options/retirement age, implementation time	Balancing delayed until 2050	Balancing delayed until 2100	Fully funded	Actuarially balanced annually
a/57, 2022	6.0%	9.3%	0.4%	4.9%
b/60, 2022	4.3%	6.7%	0.2%	7.2%
c/65, 2022	1.0%	1.6%	**-0.3%**	10.0%
d/57, 2032	2.9%	6.2%	2.4%	1.8%
e/60, 2032	2.3%	4.6%	2.5%	5.1%
f/65, 2032	0.6%	1.2%	2.5%	9.6%

alternative financing schemes is to keep the PAYG system and to delay the financial balancing to 2100, as it carries the highest value of equivalent variation. For reform options b, c, e and f, under which the retirement age is either 60 or 65 when the reform is implemented in either 2022 or 2032, the best plan among the four alternative financing schemes is to convert to the annually actuarially balanced PAYG system. The fully funded financing plan is always dominated by some other financing plans in terms of the value of equivalent variation, and it is therefore never the best financing plan for the six retirement age reform options under the optimistic fertility scenario.

3.3.3 Simulation results under the pessimistic fertility scenario

Under the pessimistic fertility scenario, the six retirement age reform options yield different values of LTSR, as reported in Table 3.5. Similar to the baseline and optimistic fertility scenarios, the LTSR increases with the retirement age, as expected. In the pessimistic fertility scenario, the financial balancing of the pension system carried out in 2022 with a retirement age of 57 results in an LTSR of 22.8%. The LTSR can be increased to 29.6% and 43.5%, if the retirement age is delayed to 60 and 65, respectively. When the financial balancing of the pension system is carried out in 2032 concurrently with the implementation of retirement age reform, the LTSR are 17.8%, 25.9% and 42.1% correspondingly for the retirement age of 57, 60 and 65. With the same retirement age, the ten-year delay in implementation of the financial balancing and retirement age reform decreases the LTSR by 5.0, 3.7 and 1.4 percentage points correspondingly for the retirement age of 57, 60 and 65. So when the retirement age is 57 or 60, the delay by a decade in launching the reform causes substantial drops in the LTSR. When the retirement age is 65, delaying reform by a decade causes a relatively small reduction in the LTSR. Compared to the baseline fertility scenario, unsurprisingly, lower values of LTSR are obtained under the pessimistic fertility scenario, as the LTSR is on average 8.5 percentage points lower across the six retirement age reform options.

Table 3.5 LTSR for the pension system under different retirement age reform options under pessimistic fertility scenario

Reform options	Retirement age	Implementation time	LTSR
a	57	2022	22.8%
b	60	2022	29.6%
c	65	2022	43.5%
d	57	2032	17.8%
e	60	2032	25.9%
f	65	2032	42.1%

Table 3.6 Comparison of welfare gains (equivalent variation) of pension financing plans under different retirement age reform options under pessimistic fertility scenario

Retirement age reform options/ retirement age, implementation time	Balancing delayed until 2050	Balancing delayed until 2100	Fully funded	Actuarially balanced annually
a/57, 2022	8.0%	13.1%	0.6%	5.8%
b/60, 2022	6.4%	10.6%	0.5%	8.4%
c/65, 2022	3.0%	4.9%	0.2%	11.3%
d/57, 2032	3.9%	8.8%	1.6%	1.8%
e/60, 2032	3.4%	7.5%	2.0%	5.3%
f/65, 2032	1.8%	3.7%	2.3%	10.1%

As under the baseline and optimistic fertility scenarios, the welfare effects across generations of the four alternative financing plans are summarised by the statistic equivalent variation, which is reported in Table 3.6 under each retirement age reform option. For reform options a, b, d and e, under which the retirement age is either 57 or 60 when the reform is implemented in either 2022 or 2032, the best plan among the four alternative financing schemes is to keep the PAYG system and to delay the financial balancing to 2100, as it carries the highest value of equivalent variation. For reform options c and f, under which the retirement age is 65 when the reform is implemented in either 2022 or 2032, the best plan among the four alternative financing schemes is to convert to the annually actuarially balanced PAYG system. The fully funded financing plan is always dominated by some form of PAYG systems in terms of the value of equivalent variation, and it is therefore never the best financing plan for the six retirement age reform options under the pessimistic fertility scenario.

3.3.4 Discussion

When comparing the projected population dynamics associated with the three fertility scenarios, it is worth noting that the optimistic fertility scenario can only lead to limited improvement in terms of the old–age dependency ratio. This shows that population ageing is bound to occur in China in the coming decades, even if the total fertility rate in China can be raised to the replacement level in only ten years as boldly assumed in the optimistic fertility scenario. The limited positive impact on the financial health of the pension system that can be brought about by the optimistic fertility scenario is reflected in the mild improvement of the LTSR, which is on average 2.6 percentage points higher under the optimistic fertility scenario across the six retirement age reform options as compared to the baseline fertility scenario.[11] It indicates that there is a limited prospect for a favourable change in the total fertility rate in China to lessen the financial burden exerted by the demographic trends on the pension system.

The comparison between the baseline and pessimistic fertility scenarios, however, reveals that a persistently low total fertility rate can present a serious challenge to the long-term sustainability of the pension system. The old-age dependency ratio reaches much higher levels under the pessimistic fertility scenario than under the baseline scenario, causing a substantial decrease in the LTSR of 8.5 percentage points on average across the six retirement age reform options.[12] Such drop in the LTSR can translate into a more than one third reduction in pension benefits for retirees, as in the case of reform option d when the retirement age remains unchanged at 57 and the financial balancing is implemented in 2032. Thus, an asymmetry is noticeable in the effects of the demographic changes in the potential for improving or worsening the financial health of the pension system. As long as the total fertility rate is of concern, the optimistic scenario does not bring about much relief to the financial health of the pension system, whereas the pessimistic scenario delivers a clear warning signal for a potential funding crisis, for which the policy makers should be prepared.

Under each fertility scenario, depending on the choice of retirement age (57, 60 or 65) and the reform implementation time (2022 or 2032), the best financing method among the four alternative financing plans can be either keeping the PAYG system and delaying the financial balancing to 2100 or switching to an annually actuarially balanced PAYG system. Both plans generate tangible welfare gains for people across the generations as measured in equivalent variation in their lifetime consumption, as compared to the benchmark financing method in each reform option, which is based on PAYG with financial balancing in either 2022 or 2032 (i.e. reform options a, b, c, d, e and f themselves being used as the benchmark to compare the alternative financing plans). The fully funded system is *never* the best financing plan in *all* fertility scenarios and under *all* reform options. It therefore indicates that the best choice of financing methods for the pension system in China remains a PAYG system. However, the parametric details of the PAYG system ensuring welfare optimality, e.g. when to implement the financial balancing of the pension system and whether to make the system annually actuarially balanced, vary with the retirement age reform options, which suggests a feature of path dependence that should be taken into account in the policy making process for the new pension system design in China.

The timing of launching the retirement age delay has some impact on the affordable pension benefits level. In general, postponing the reform by ten years (i.e. 2032 vs. 2022) decreases the LTSR. Under the baseline and optimistic fertility scenarios, the reductions in the LTSR are relatively small (reductions all below two percentage points) and minor for the case of retirement at 65 (0.3 and 0.2 percentage points for the baseline and optimistic scenarios, respectively). Under the pessimistic fertility scenario, the ten-year delay in activating the reform is more costly as manifested in the larger reductions in the LTSR (5.0 percentage points for retirement at 57, 3.7 for 60 and 1.4 for 65). Thus, it would be a cautious measure for the policy makers to closely monitor the fertility level and its trends in China before deciding when to implement the retirement age reform. Postponing the retirement age to 65 seems to be able to mitigate some of the negative financial impact of introducing the reform at a later time, as the ten-year delay in implementation of the reform only causes relatively small reductions in the LTSR in each of the three fertility scenarios if retirement starts at 65.[13]

To determine whether there exists a fertility cliff under which the pension system will be in de facto bankruptcy with all plausible retirement age reform options, it is necessary to examine whether any reform option is powerful enough to counter the negative financial impact of the worst-case scenario. More specifically, it can be determined by looking at the LTSR achievable by the highest viable retirement age (i.e. 65) in the pessimistic fertility scenario and comparing the LTSR with a standard as a threshold for defining old-age pension adequacy. Postponing retirement to 65 can achieve an LTSR of 43.5% if implemented in 2022 or 42.1% if implemented in 2032, under the pessimistic fertility scenario. The ILO has over the years established a number of conventions on pension provision, which are agreed upon by a wide range of governments, employers and trade unions. A replacement ratio of 45% is set by Convention 128 of the ILO to "guarantee protected persons who have reached a certain age the means of a decent standard of living for the rest of their life" (Humblet and Silva 2002).

The highest LTSR achievable in the pessimistic fertility scenario is 43.5%, slightly below the replacement ratio of 45% as set by the ILO as a threshold to protect the decency of life for the retirees. Therefore, the possibility of the fertility cliff for China cannot be ruled out. Such results should serve as a reminder to the government that they should pay more attention to monitoring the fertility rate in China and its future trends in a more timely, transparent and accurate manner. Should the fertility rate drop even further than its current level, the government should take more proactive measures to reverse the adverse trends in fertility in order to avoid a funding disaster on the pension system in the future.

3.4 Conclusion

The results of the quantitative simulation of the pension system in China under the three fertility scenarios demonstrate an asymmetry in the effects of the demographic changes on the financial health of the pension system. Even a very optimistic fertility assumption cannot improve the future profile of the old-age dependency ratio by much compared to the baseline fertility assumption. However, a persistent and very low level of fertility rate, which could happen in a country that has fallen into the low fertility trap, can substantially increase the future old-age dependency ratio and thus cause considerable financial pressure on the pension system.

Under each retirement age reform option and in all fertility scenarios, some form of PAYG system turns out to be the optimal financing plan that dominates the fully funded system in terms of welfare gains across the generations. The timing of launching the reform to postpone the retirement age has some effect on the long-term sustainable pension benefits level. Delaying the reform by ten years lowers the LTSR in general, and such an effect is magnified under the pessimistic fertility scenario.

As the highest LTSR achievable under the worst-case scenario in the quantitative simulation is below the threshold set by the ILO for defining pension adequacy, the possibility of a fertility cliff cannot be ruled out. The government should closely watch the change of future fertility rate in China and adopt pronatalist policies to help increase the birth rate level when it falls into the direction of the fertility cliff that would force the pension system in China into de facto bankruptcy in the decades to come.

The results of this chapter refute several existing studies that claim increasing retirement age is sufficient for solving the funding crisis of the pension system in China (Zeng 2011; Zhongmei Yuan 2013; Song et al. 2015), while affirming some other studies that find the retirement age reform to be only a necessary ingredient for solving the fiscal problem posed by the fast ageing population (Yuan 2014; Yu and Zeng 2015; Yang and He 2016). The findings of this chapter enrich the current debate on the impact of population ageing on pension sustainability in China by assessing the financial sustainability of the system based on its ability to avoid de facto bankruptcy rather than focusing on the nominal sense of bankruptcy that could be always prevented with parametric adjustments in contribution rate, benefits level or retirement age. The possibility of the fertility cliff is identified with a wider range of fertility scenarios than typically considered, which is an area of analysis underexplored in the existing scholarship.

The results of the analysis conducted in this chapter demonstrate that, abstracting from the impact of the fragmented nature of the pension system, reform to the retirement age alone can only provide a necessary but not a sufficient condition for ensuring the system's long-run financial sustainability. As will be shown in the next two chapters, the fragmentation of the system generates significant negative effects on pension sustainability, which reinforces the conclusion about the necessity but not the sufficiency of the retirement age reform for ensuring pension sustainability in China.

Notes

1 As the legal minimum working age is 16 in China according to the Labour Law (National People's Congress 1994), people aged below 16 are counted as young dependents. People aged 60 and above are counted as elderly dependents, taking into account the current average retirement age of around 57 (Song et al. 2015) and the likely postponement of retirement age in the coming decades. These two groups of population are projected to account for 14.1% and 34.6% of the total population, respectively, in 2050 (United Nations 2019a), leading to a ratio of young and elderly dependents in the total population of 48.7%.

2 According to the low fertility trap hypothesis, when the total fertility of a country falls below 1.5, it is conjectured that a self-reinforcing mechanism driven by demographic, sociological and economic factors will generate a downward spiral in future fertility (Lutz, Skirbekk, and Testa 2006; Jin 2014).

3 Anderson and Kohler identify high private education costs and a highly competitive job market that prizes academic achievement in East Asian countries such as Korea as a key factor contributing to the very low fertility rate in these countries, as couples prefer quality over quantity in their reproductive and child-rearing decision making and can only afford having no more than two children in most cases (Anderson and Kohler 2013). The very costly private education in these economies is confirmed by a recent report, in which China, Taiwan, Hong Kong and Singapore are top-ranked in terms of average education expenditures in the world (HSBC 2017).

4 A prominent example of countries having fallen into the low fertility trap is Japan, where the fertility rate has been following a largely downward trend since 1950 and below the replacement level for the past 60 years (United Nations 2019c).

5 The replacement level of total fertility rate is about 2.1 on average for a typical population. The exact value of the replacement level of total fertility rate differs slightly from country to county depending on their particular demographic parameters such as the mortality rate distribution across gender and age groups as well as the size of the age cohorts in their populations. The value of the replacement level of total fertility rate for China is calculated to be 2.0763 in the model.

6 According to the "Main Data Bulletin of the Sixth National Population Census in 2010", the total fertility rate in China was 1.181 (National Bureau of Statistics 2011). The further decline in fertility levels in China in recent years is witnessed by these official figures of the total fertility rate.

7 The assumption of a total fertility rate at 0.9 for the pessimistic scenario is very low but not entirely unrealistic. For example, the total fertility rate in South Korea reached 0.98 in 2018 (The Economist 2019).

8 No extra injection of government funds is allowed besides the initial pension fund in order to explore the long-term financial sustainability of the pension system as a self-financing system. For more technical detail of the definition of the financial balancing the pension system, refer to Equation (11) in Song et al. (2015).

9 The reform options a and d both have the retirement age of 57, which is equal to the current average retirement age and serves as the default reform option and also as a reference point for comparing the effects of postponing the retirement age.

10 For details of the mathematical definition of the welfare criterion of the social planner, refer to Equation (12) in Song et al. (2015).

11 See Tables 3.1 and 3.3.

12 See Tables 3.1 and 3.5.

13 See Tables 3.1, 3.3 and 3.5 for results related to the LTSR as discussed in the paragraph.

4 Fragmented administration and pension sustainability in China

This chapter studies the impact of the fragmentation in pension administration and financing on pension sustainability by identifying and analysing two mechanisms through which the fragmentation affects pension sustainability: (1) moral hazard and (2) inefficiencies. The first section reviews the administrative and public financial environments the Chinese pension system operates in. Section 4.2 examines the intergovernmental relations in pension financing and how they lead to a principal-agent problem, from which the moral hazard arises. Section 4.3 offers an in-depth analysis of the behavioural patterns of local governments under moral hazard and documents the detrimental effects of their manipulations of the local pension pools with detailed examples from case studies. Section 4.4 investigates inefficiencies caused by the fragmentation that have direct negative impact on the financial sustainability of the system. Section 4.5 concludes the chapter by discussing on the aggregate effect of local implementation of pension policy on the financial sustainability of the pension system.

4.1 Administrative and public financial environments for pension services provision

The development of the pension system in China is shaped by the characteristics of the country's administrative and public finance regimes. The relationship and interaction between different levels of government in China are dynamic and complex, and they define the institutional environments for the functioning of the pension system.

The Chinese public administration system is hierarchical, comprising of five levels of government. Under the central government, there are 33 provincial-level governments (including the two special administrative regions of Hong Kong and Macau), 334 prefectural-level governments, 2,851 county-level governments and 39,862 township-level governments (National Bureau of Statistics 2017). The central government only employs less than 5% of all the public employees in China, making it small relative to the local governments in terms of number of staff (Wong 2018a). It is also unusually small compared to other large countries such as the US, the UK, France, Japan and India, with the proportion of public employees hired by the central government in these countries ranging from 16% to 66% (Zhou 2010; Wong 2018a).

DOI: 10.4324/9781003182696-4

With such a small central government at the top of the state bureaucracy in China, the implementation of policies and the delivery of public services, including those most important and expensive services such as education, health care and social security, are largely delegated to the subnational governments at lower levels, which leads to a high degree of decentralisation of the fiscal responsibility for public expenditures (Wong 2010). This is reflected in the share of local governments in public expenditures in China. In the period from 2000 to 2011, the local governments accounted for 71% of total public expenditures; the share of the local governments was even higher at 86% for social expenditures including health care and pension (National Bureau of Statistics 2012). From 1993 to 2009, the share of the central government in budget expenditures decreased from 34% to 20%, while the share of the lower-level governments at county and township levels kept increasing from 27% to 40%. However, the decentralised expenditure responsibility has not been matched on the revenue side. Since the 1994 tax sharing reform, the share of local governments in total budget revenues hovered below or around 50%, with the central government taking the lion's share of the revenues in most years (National Bureau of Statistics 2019a).

Such mismatch in intergovernmental relationships in expenditures and revenues is one of the hallmarks of the lopsided intergovernmental fiscal relationship in China. And the distortions it created have been identified as major factors contributing to inefficiencies in the Chinese economy (World Bank 2002, 2007; Wong 2018b). It has led to the heavy reliance of subsidies and transfer payments for local governments to finance their expenditures. With the increase in the share of local governments in budget expenditures and the proliferation of programmes aimed at improving public services and expanding the social safety net since the mid-2000s under the rubric of "creating a harmonious society", transfers from the central government to local governments increased tremendously. For these growing transfers to reach the counties and cities, they have to pass through each level of government, creating great administrative burden on the bureaucracy that outstrips the capacities of the monitoring and supervisory systems, while generating many undesirable side effects such as cost inflation and wastefulness (Wong 2016, 2018c).

The high degree of administrative decentralisation is also reflected in how relationships between bureaucracies are configured in China. As noted by Saich, a bureaucratic unit of a local government in China is answerable both vertically to its counterpart unit at higher levels and horizontally to the political leaders of the jurisdiction where it is based, resulting in a dual-leadership arrangement in the bureaucratic coordination system (Saich 2004). Lieberthal characterises the Chinese bureaucratic system as one of "fragmented authoritarianism", where "officials of any given office have a number of bosses in different places". He points out that only one of these bosses can be the primary leader and that "the primary leadership over a particular department resides either on the vertical line (tiao) or the horizontal piece (kuai)" (Lieberthal 2004).

The local governments are capable of materially influencing the policy implementation process in their jurisdictions in most aspects of the national socioeconomic policy spectrum including pension policy, because they have primary leadership over most of the government departments and agencies of the same administrative level and control the personnel and budgetary allocations for these units that are in charge of implementing the policies on the ground. In other words, the local branches of most government bureaucracies that carry out specific functions such as pension administration are obliged to place higher priority on the concerns of the local government of the same county or city than on those of their functional counterparts at higher administrative levels. For example, as a county-level government is responsible for any remaining deficit in the local pension pool after receiving subsidies from higher levels, it is also able to manage the behaviour of the county social security bureau by controlling the disbursement of government funds for personnel, budgets and properties (ren-cai-wu) for the bureau.

Compared to the international experiences, fragmentation in administration and financing is a unique and salient feature of the Chinese pension system. But the highly decentralised management of the pension system is unexceptional in the context of public services provision regime in China: the duties for providing almost all public services including education, healthcare and social security are delegated to lower-level governments often at the county level (Wong 2013b). However, the decentralised and fragmented administrative structure for pension policy implementation has implications for the performance of the pension system. With the system divided into over 2,000 pools mainly managed at city or county level (B. Zheng 2012), the enrolment size for each pension pool ranges from 13 million in Beijing and 11 million in Shanghai (National Bureau of Statistics 2018a) to below 30,000 in some county level cities.[1] The performance of the pension system as a whole therefore depends on the aggregate effect of how each of the individual pension pool is functioning. In order to analyse the impact of fragmentation on the performance of the system and its financial sustainability in particular, it is thus important to examine how the local governments would behave in their role as the provider of pension services in their jurisdictions with the constraints and incentives embedded in the administrative and public financial environments.

Figure 4.1 provides a schematic summary of the institutional environments of pension policy implementation. The intergovernmental relations underpin the institutional settings for the policy implementation environments. The central government provides policy design and a legal framework, and it delegates the administration of the pension system and the implementation of pension policy to provincial-level governments, which determine key parameters of the pension system in each province including the contribution rates on employers. A provincial government then further delegates the implementation of policies and the administration of the pension system to lower-level governments. The county- or city-level governments implement pension policy on the ground and do the heavy lifting of managing the local pension pools and providing pension services for the local population in their jurisdictions. While the pension policy and the relevant legal framework are determined by the central government, local governments at the provincial, prefectural[2] and county levels exercise considerable level of discretion to interpret the national policy and decide on the details for how it should be implemented in the local context. This inevitably leads to variability in pension policy outcomes, while the aggregate effect of the local policy implementation and

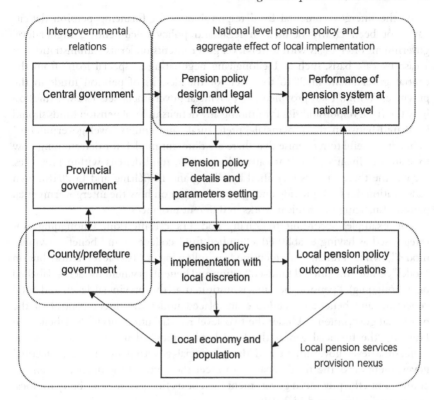

Figure 4.1 Institutional environments for pension policy implementation.

outcome variations across some 2,000 pension pools determines the overall performance of the pension system including its financial sustainability. The county- or city-level governments are the provider of pension services in their jurisdictions. Whether they will fulfil their fiduciary duty as the custodian of local pension funds and comply with the spirit of the pension policy and relevant legal provisions set by the central and higher-level governments will have material impact on the performance of the pension system.

4.2 Intergovernmental relations in pension financing

The central government is in charge of designing the national pension policy and stipulating the relevant legal provisions. But due to the fragmentation of the system with large variations in the details of pension policies across regions, the legislation on pension policies only serves as a legal framework to set broad guidelines. The Social Insurance Law only provides some principles on the obligation for paying contributions on employers and employees and acknowledges the government's responsibility in providing subsidies, without clearly spelled rules on how the government will allocate subsidies to pension funds of different provinces or how fiscal responsibilities should be shared across different levels of government.[3]

As the intergovernmental fiscal responsibilities for financing pension deficits have not been clearly defined in legal and policy documents, a higher-level government often deals with lower-level governments under its administration on a case-by-case basis, such as Heilongjiang negotiating a "special loan" from the central government in 2016 to cover for the deficits of pension funds in the province.[4] Within each province, the provincial government sets its own rules on how the fiscal responsibilities for funding any deficits in local pension funds should be shared between the provincial-, prefectural- and county-level governments.[5] While it is useful to recognise that there are different models for determining how to share the financial burdens caused by pension fund deficits within provinces, such as the three models identified by Zhang and Li (Zhang and Li 2018), such classification does not provide enough information on how the intergovernmental pension financing mechanism works in different provinces.

For example, according to Zhang and Li's classification, Chongqing was categorised as having established a "unified fund-collection and benefit-payment model", while Hunan was identified as following a "two-level redistribution model". Under the unified fund-collection and benefit-payment model (as labelled on Chongqing), "pension pooling is instituted at the provincial level, and fund collection and benefit expenditure are placed under the direct control of the provincial government". Under the two-level redistribution model (as labelled on Hunan), "the financial gap at the local level is plugged using two sources: the prefectural-level government and the provincial government. If the prefectural government lacks the fiscal capacity to meet the demand for financing pension entitlements, the provincial government then bears financial responsibility for local pensioners" (Zhang and Li 2018).

According to a deputy director in the municipal bureau of human resources and social security, Chongqing was actually following a "differential fiscal responsibility sharing mechanism" between the municipal government and a county- or district-level government for funding the pension fund deficits, which treats each county or district in Chongqing differently depending on how the targets for pension revenues and expenditures are met each year (Xinhua Net 2019b). And according to a pension policy document issued by the government of Hunan in 2019, the provincial government decides on how it should share the fiscal responsibilities with prefectural- and county-level governments in Hunan "based on factors of the local economy and society (in each county) such as medium-to-long-term socioeconomic development planning, economic development status, demographic structure, employment situation, fiscal capacity, number of participants in the pension insurance schemes and dependency ratio" (People's Government of Hunan 2019). As every county or prefecture is different in terms of such a comprehensive set of local contextual variables, the county or prefectural government can always claim that they face a unique situation and should be treated differently on a case-by-case basis by the higher-level government when negotiating how the financial responsibilities for pension fund deficits should be shared and how much subsidies should be given.

In the mid-1990s, there were widespread cases of retirees having their pension benefits in arrears, contributing to the rising social unrest in that period (Zhu 1998;

Tanner 2005). Since 1998, the central government launched the "two assurances policy" to ensure, firstly, that the basic livelihood of laid-off workers from state-owned enterprises be guaranteed, and, secondly, that retirees receive benefits on time and in full amount (National People's Congress 2010b). From 1998 to 2001, to help implement the "two assurances policy", the central government provided
· 130 billion yuan of earmarked transfer payment to the old industrial bases and the central and western regions (Xinhua Net 2002). Such assurance policy for retirees was reiterated by the State Council in 2005 (State Council 2005). Since then, no large-scale pension arrears happened, as fiscal subsidies allocated to help finance the UES increased substantially since 2005, increasing by three fold in five years from 65.1 billion yuan in 2005 to 195.4 billion yuan in 2010 and then having quadrupled by 2017 to reach 800.4 billion yuan or 21.4% of total annual pension expenditures of the UES (Ministry of Labour and Social Security 2006; Ministry of Human Resources and Social Security 2011a, 2018).

As pointed out by Lou Jiwei, the just-retired Chairman of the National Social Security Fund and former Minister of Finance, the rapid growth in subsidies to the UES indicates that the pension system is ridden with too many defects that are caused by the moral hazard arising from decentralised and fragmented management (Lou 2019). With huge amount of subsidies to bail out the local pension pools in deficits and the importance the central government has placed on ensuring the timely and full disbursement of pension benefits in all places, the local governments effectively face a soft budget constraint on their local pension funds, as they have always managed to get enough amount of subsidies for more than a decade now from higher-level governments, with the central government being the ultimate guarantor of the solvency of the pension system and the main source of subsidies. The county-level governments have formed the expectation that they will be bailed out by higher-level governments when their local pension funds are in deficit. As explained by an official interviewed in a county, ensuring the timely disbursement of pension benefits is a "political task (zhengzhi renwu)" assigned by higher levels, which has to be carried out to ensure social stability regardless of its financial cost.[6] The unclear division of responsibilities between levels of government in China in pension financing and deficit settlement has created a principal-agent problem, where moral hazard arises to compromise the financial sustainability of the pension system (Wong and Yuan 2020).

As a latest move in the reform on pension financing, the State Council issued the "Notice of the State Council on the Establishment of the Central Adjustment System of the Basic Pension Insurance Fund for Enterprise Employees" to set up a Central Adjustment Fund (CAF) (State Council 2018). The CAF started to operate in July 2018 with clear rules: each province contributes to the CAF a percentage of the product of the average wage in the province and the number of working-age participants of the UES in the province,[7] and each province receives from the CAF an amount that is proportional to the number of retirees of the UES in the province; the total contributions to the CAF are fully redistributed to each province according to the above rules every year. While the CAF works as a risk-sharing mechanism across provinces with much more clearly articulated rules and is seen as a major step towards achieving a nationally unified pension system, the problems associated with the moral hazard will not be resolved without further reform. In particular, it remains unclear how the redistributed funds from the CAF together with other subsidies from the central government will be allocated by the provincial governments to the prefectures and counties within each province.

The examples of Chongqing and Hunan mentioned previously reflect the still ad hoc and case-by-case nature of allocation of subsidies for the UES in these two provinces in 2019, which was after the central government's decision to set up the CAF. The county or prefectural governments are often given annual quotas for pension contributions and expenditures by a provincial government as a way to ensure that local pension pools are managed with a certain level of care. These examples, together with the information obtained from fieldwork, indicate that the financing mechanism of the system often falls back to a form of contracting that is highly reminiscent of the revenue-sharing system before the tax sharing system reform was implemented in 1994 (Wong and Yuan 2020).[8] But the moral hazard arises, when the local governments are somehow able to obtain enough subsidies from higher levels to keep the local pension pools afloat, even if they fail to achieve those quotas, leading to a soft budget constraint.

Such features of the intergovernmental relations in pension financing are symptoms of a deeper problem that lies in the dysfunctional intergovernmental fiscal relationship in China. They lead to a murky and complicated negotiation process between levels of government on pension fund deficit settlement[9] and a soft budget constraint for county- or city-level governments that implement pension policies and manage local pension pools in their jurisdictions, setting the stage for these lower-level governments to act with discretion or even outright noncompliance to national pension policy set by the central government with seeming impunity.

4.3 Moral hazard and behavioural patterns of local governments

The moral hazard caused by the disarticulated intergovernmental fiscal responsibility in pension financing and deficit settlement poses a challenge for ensuring the long-term sustainability of the pension system. Given the soft budget constraint on the local pension pools and an incentive structure that rewards the local political leaders for achieving fast growth of the local economy, the local governments behave in ways that incur systematic implementation bias to the detriment of the financial sustainability of the pension system as a whole. The compliance to rules concerning the collection of pension contributions, the criteria for enrolment and benefits entitlement as well as the financial management of future liabilities is particularly important. The incentive structure and behavioural patterns of the local governments in these aspects of pension policy implementation and compliance are studied in detail in this section.

4.3.1 Noncompliance to pension contribution rules

The local governments lack the incentive to fully comply with the national pension policy on rules regarding pension contributions and their collection, and they are not subjected to sufficient institutional constraints when they break these rules. This has led to widespread underpayment and evasion of pension contributions by both employers and employees, which poses a serious challenge to pension sustainability from the revenue side of financial management of the pension system.

Under the central government policy, certain rules are applicable to the calculation of the pension contributions payable by employers and employees.[10] For the contributions payable by employees, there is a permissible salary range for the calculation. According to the official interpretation of the Social Insurance Law, "employees with salaries lower than 60% of the local average wage must pay contributions calculated using 60% of the local average wage as the base salary. Employees with salaries higher than 300% of the local average wage must pay contributions calculated using 300% of the local average wage as the base salary" (Ministry of Human Resources and Social Security 2012). The practice of setting a permissible salary range for calculating employees' contributions was adopted in the mid-1990s (State Council 1995). The lower limit at 60% of the local average wage was meant to discourage firms from reporting artificially low wage figures (West 1999).

For the contributions payable by each employer, the central government policy is clear that they should pay according to the actual amount of total wage bills of the firm. According to Item 3 of the Document 26 issued by the State Council in 1997, the initial policy document to launch the UES nationwide,

the pension contribution rates payable by an enterprise (hereinafter referred to as enterprise contributions) shall generally not exceed 20% of the total wages of the enterprise (including the part allocated to individual accounts), and the specific rate shall be determined by the people's governments of the provinces, autonomous regions or municipalities directly under the central government.

(State Council 1997)

The Social Insurance Law reiterates this principle, as Clause 12 states that "the employer must pay pension contributions into the social pooling account in proportion to the total wages of the employees hired by the employer in accordance with state regulations" (National People's Congress 2010a). The obligation of both employers and employees to participate in social insurances including the pension insurance programme and pay social insurance fees in accordance to the law is also stipulated clearly in Clause 72 of the Labour Law (National People's Congress 1994).

Yet, in practice, the contribution rules that are supposed to be applicable only to the employees' contributions came to be applied to the employers' contributions as well. Moreover, the wrong set of rules has been applied on employers in a "wrong" way as well, as local governments across the country have allowed employers to calculate their pension contributions according to the imputed wage, often at 60% of the local average wage, instead of the actual payrolls. Even for the employees' contributions, the rules by the central government allow the imputed wage at 60% of local average wage to be used only when an employee's actual wage is below 60% of local average wage; when the actual wage is within the range from

60% to 300% of local average wage, the actual wage must be used; when the actual wage is above 300% of local average wage, 300% of local average wage must be used (Ministry of Human Resources and Social Security 2012).

According to the local officials interviewed in several provinces, it is common for private companies to make pension contributions using the imputed wage at 60% of local average wage when the actual wages of their employees are higher than 60% of local average wage, while most of the large listed companies and state-owned enterprises follow the rules and pay contributions according to actual payrolls.[11] The higher compliance rate among these large or state-owned enterprises is unlikely the result of tighter enforcement by local governments on them. Rather, it is the rational choice for most of these enterprises to self-report and pay the right amount of pension contributions: big listed companies have their reputation at stake if they are found to act illegally on labour issues such as evading social insurance fees, while state-owned enterprises have different incentives than private companies and do not need to compromise the welfare of their employees and risk breaking the law to maximise the profits.[12]

One survey in 2018 found that only 27% of firms in the sample fully complied with the Social Insurance Law in making pension contributions, while 32% of firms made contributions using the minimum permissible salary (51Shebao 2018), which was 60% of the local average wage in the previous year in most areas. According to an entrepreneur in Shanghai, none of the local small-to-medium sized private firms were paying pension contributions according to their actual payrolls to his knowledge. His own firm paid pension contributions for all of its some 30 employees including himself as the managing director according to 60% of the local average wage. "Social insurance fees weigh heavy on our company," he remarked, "because it is a major component of labour costs. If we don't minimise the amount, other companies will. And we will be out of business soon."[13] As private firms had accounted for over 80% of urban employment in China by 2018 (Xinhua Net 2018a), the unwillingness of local governments to enforce pension contribution rules on them implies that the self-financing ability of the pension system is seriously handicapped.[14]

Allowing the firms to pay pension contributions according to 60% of the local average wage is equivalent to giving them a 40% "discount" on the contributions for the employers on average. A local government will be strongly discouraged to follow the central government policy strictly on pension contribution rules in its jurisdiction, if any other local governments choose to allow their local firms to enjoy the discount on pension contributions. To unilaterally obey the rules will lead to a less competitive business environment and reduce the attractiveness to investment for the local economy, a price the local leaders do not want to pay. The result is the widespread acquiescence of local governments to the rule-defying practice by most private companies to report and pay pension contributions according to the imputed wage rather than actual payrolls, a perverse equilibrium achieved during the race to the bottom.

The local governments also set up "local policies (difang zhengce)", some of which are tantamount to outright noncompliance to the central government policy, to further interfere with the implementation of pension policy on contribution rules, resulting in different sets of rules across provinces and cities and sometimes even for different firms within the same city.[15] For example, the lower limit of the permissible salary range can be even lower than 60% of local average wage in some places. In Beijing in 2018, there were five tiers of the base salary for calculating pension contributions, which were at 40%, 60%, 70%, 100% and 300% of the local average wage, applicable to different types of enterprises; in 2019, the lowest tier in Beijing was increased to 46% from 40% of local average wage (Bureau of Human Resources and Social Security of Beijing 2018; Beijing Municipal Social Insurance Funds Management Centre 2019). No reason was given for the increase in the lower limit for salary base in Beijing in the local policy documents. But according to an article released on Xinhua Net, the increase in the lower limit will continue to reach 52% in 2020 and 60% in 2021, and such increase accompanies the decrease in contribution rate on employers from 20% to 16% starting from May 2019 (Xinhua Net 2019a). In Xiamen of Fujian, the lower limit was 27% of local average wage in 2018 (Bureau of Human Resources and Social Security of Xiamen 2018).

These cities with lower-than-usual levels of the lower limit of the permissible salary range tend to be economically vibrant. With a relatively young workforce and therefore a healthier dependency ratio for the local pension system, such a city typically runs a surplus in the local pension fund and does not need to rely on transfer payment from higher-level governments to fund the pension related expenditures. Consequently, they have higher bargaining power in deciding how to manage their local pension system, as indicated by their ability to defy the regulations of the central government and the official interpretation of the Social Insurance Law regarding the pension contribution rules in this case. Besides giving the local firms an extra edge with an even lower level of pension contributions, there are other motives behind the decision by some local governments to set the lower limit below the usual level of 60% of local average wage, which are discussed later in the second case study in Section 4.3.2. As a result of all these distortions and interferences, the actual contribution rate (contributions as percentage of wages) differs significantly across provinces, from 8.4% in Guangdong to 27.3% in Liaoning in 2015 with the national average rate at 18.0%, which were all below the national guideline rate of 28% (National Bureau of Statistics 2016a).[16]

Besides decreasing the lower limit for the permissible salary range, the local governments are able to interpret the definition of the local average wage in ways they see fit. For example, local government in some regions can use either the arithmetic mean or the median value for local wages as the local average wage to be used to calculate pension contributions and benefits in a particular year.[17] The local average wage used for calculating pension contributions can also be different from the one used for calculating pension benefits. In a central province, for calculating pension benefits, the local average wage was based on the figure in the previous year, while the local average wage for calculating pension contributions had been fixed at the local average wage in 2016 for the three consecutive years from 2017 to 2019, giving further discount on pension contributions for companies in the province.[18]

The local governments are also reluctant to take real efforts to ensure full collection of pension contributions on the local firms, even after the employers' portion of pension contributions has already been reduced through the measures to fiddle with the system as described above. This is reflected in the lack of manpower allocated by the local governments to carry out enforcement and supervision on firms (Tan and Tan 2016). In 2019, in a county with a population of 750,000 people and several thousand companies, only three staff in the county bureau of human resources and social security were assigned to verify the data reported by companies for calculating the pension contributions, and they also had other duties such as checking on data related to the disbursement of pension benefits in the county.[19] Combined with the lack of incentive for private firms to comply with the Social Insurance Law, this resulted in widespread underpayment and evasion of pension contributions.

4.3.2 Silent inflation of future liabilities

As shown in the analysis above, the noncompliance to pension contribution rules damages pension sustainability by reducing the self-financing capacity of the system. But the negative effects of the moral hazard are not limited to the revenue side of the financial management of the pension system. The local governments also manipulate other parameters of the system, including the benefits level and the criteria for enrolment, which has a more insidious impact on pension sustainability by silently inflating the future liabilities of the system.

Allowing early retirement and increasing benefits level are two ways the local governments have been noted to take to increase the level of future liabilities of the pension pools. For example, following the two rounds of notice for pushing for higher-level pension pooling released by the State Council, there were correspondingly two waves of early retirement in 1990s, as the local governments tried to maximise the claim of the local populace for future pension benefits while anticipating the higher-level governments to pay for the costs (Xia 2001). A policy document issued by the central government indicates that some local governments raised the pension benefits level in their jurisdictions in the early 2000s, despite the objection of the central government. In the "Notice of the General Office of the State Council on the Prohibition of Local Governments from Raising the Level of Benefits for the Basic Pension Insurance for Enterprises on Their Own" issued by the State Council in 2001, it is mentioned that, despite the exhortations and demands made by the State Council in 2000:

> There are still some places where the local governments have raised the level of basic pension benefits for retirees on their own without the consent of the State Council, and some places have raised the benefits level blindly even when their basic pension insurance funds are in deficit. Such practice not only undermines the authority and unity of the relevant national policies, but also increases the difficulty of ensuring that basic pensions are paid in full and on time. At the same time, it has caused cross-regional comparisons and is not conducive to social stability and must be stopped.
>
> (State Council 2001)

In some rich regions with pension fund surpluses, such practice of introducing higher-than-normal level of increases in pension benefits for the local retirees continues, with a newly coined phrase "welfare championship (minsheng jinbiao-sai)" associated with it.[20]

As the number of retirees increases with the ageing population, an increase in the benefits level not only requires an increase in the pension expenditures for the current year, but it also leads to increased expenditures in the coming years and decades as it adds to the base level from which future increases in benefits will be added on. However, the full financial impact of such increases in the benefits level is not captured by the accounting method adopted for reporting the budget for social insurance funds. Despite some progress made recently for establishing a government financial reporting system based on accrual accounting,[21] no material change has occurred in the disclosure of financial information regarding pension funds. The budget reports for social insurance funds including pension funds released by various levels of government are still recorded based on cash accounting, which only shows the revenues and expenditures incurred in the current year without any estimation of the value of total future liabilities. As such, the long-term impact of the manipulations on the pension pools by local governments is hidden from public oversight.

4.3.2.1 Policy on the buy-in option

The buy-in option for the UES is another policy instrument used by the local governments to manipulate the local pension pools with long-term financial implications. The buy-in option allows eligible individuals to join the UES by paying a lump sum fee for up to 15 years of pension contributions and therefore become entitled to receive pension benefits when reaching the retirement age. The Ministry of Human Resources and Social Security and the Ministry of Finance jointly issued the "Opinions on Resolving the Remaining Issues such as Basic Old-Age Social Security for Retirees of Uncovered Collective Enterprises" in 2010 (Ministry of Human Resources and Social Security and Ministry of Finance 2010). This is the first policy document issued by the central government with specific policies for former employees of collective enterprises to join the UES through buy-in. But before the release of this policy, some local governments had been allowing previously uncovered retirees of collective enterprises to join the UES through certain measures for several years, which is acknowledged in Item 2 of the policy document:

> In recent years, some regions have introduced policies for the inclusion of retirees of collective enterprises that were uninsured by the UES in accordance with local conditions, and have solved some outstanding problems. Due to the differences in the initial conditions of the work in various places, the progress of solving this problem can be fast or slow. All regions should proceed from reality, adhere to the people-oriented principle, conscientiously do a good job in the investigation and survey work required, scientifically define the scope of retirees of uninsured collective enterprises, and accurately grasp the number

of personnel that should be included in the UES; in the implementation of the policies, it is necessary to clarify specific standards, avoid causing cross-regional comparisons, formulate specific work arrangements, clarify the time limit for completing tasks, and avoid letting the problems left over from history remain unresolved for long.

Item 1 states that "retirees from some groups of collective enterprises in financial difficulties have not joined the UES in time or continued to maintain their participation in the UES and are now lacking the financial security to support their basic needs, and that such situation has become one of the prominent contradictions affecting social stability." The document sets the goal of resolving these remaining issues such as the access to pension benefits for the retirees from collective enterprises by the end of 2011. In Item 3 of the document, the eligibility for the buy-in is defined as "those who have urban hukou and have established labour relationships or formed factual labour relationships with urban collective enterprises". For those satisfying these conditions, if they had reached or exceeded the statutory retirement age before 31 December 2010, they were eligible to pay a lump sum fee for up to 15 years of the pension contributions required for the UES as the buy-in fee. If they had not reached the statutory retirement age on 31 December 2010, they must join the UES and pay the contributions according to relevant regulations. If the accumulated contribution history was to be less than 15 years when they reached the statutory retirement age later, they could pay a lump sum fee for the number of years of contributions short of the 15 years required.

Local governments issued their own policies on buy-in as early as 2003. From 2003 to 2012, 21 provinces released local policies on buy-in (Chen 2013). New policies and adjustments to their existing policies on buy-in have been made since then by local governments. The policies vary in the details in terms of the criteria for eligibility for buy-in and the method used to calculate the buy-in fees, as can be seen in several publicly available local policy documents issued by some provincial governments.[22] It is also common for local governments to suspend the implementation of buy-in from time to time. For example, Tongren, a prefectural government in Guizhou, issued a notice to inform that the implementation of buy-in will be stopped from 1 October 2018 onwards (Bureau of Social Insurances of Tongren 2019).

It is worth noting that, if the central government policy is strictly followed by the local governments, the buy-in option for people reaching the retirement age to pay 15 years of contributions in one lump sum payment and hence becoming eligible for receiving pension benefits immediately should have been a one-off phenomenon in 2010. This is because after 2010, no one should be able to pay a lump sum to cover for the entire 15 years of contributions, as this person must have joined the UES and made contributions regularly since 2010. For example, if this person reached retirement age in 2015, she or he must have accumulated five years of contributions and thus only needed to pay for ten years in a lump sum as the buy-in fee. In reality, local governments across the country have from time to time allowed their local residents by hukou to buy in by paying for 15 years of contributions after 2010.[23] Also, the eligibility criteria were loosely defined with the expression "those who have urban hukou and have established labour relationships or formed factual labour relationships with urban collective enterprises". Combined with the poor record keeping for employment history in 1970s, 1980s and even 1990s, this would give rise to many loopholes that can be exploited to allow almost anyone interested to become eligible for buy-in.

In 2016, the Ministry of Human Resources and Social Security and the Ministry of Finance issued another policy document "Notice on Further Strengthening the Management of Revenues and Expenditures of Basic Pension Insurance Funds for Enterprise Employees". In Item 1.3 of the document, it requires that "all local governments must not, in violation of the state regulations, adopt the method of charging a one-time, lump sum fee to include non-qualified personnel exceeding the statutory retirement age into the UES, and that all local governments must stop implementing the policies devised by themselves to expand the scope of eligibility for buy-in beyond the scope defined by the state immediately" (Ministry of Human Resources and Social Security and Ministry of Finance 2016a). The release of this policy document indicates that the central government realised the seriousness of the financial implications of allowing too many people to buy in and wanted to curb the behaviours of local governments in allowing excessive buy-ins by people who were not supposed to be eligible. But as shown in the following case studies, the local governments did not cease to exploit the loopholes of the policy for self-serving purposes.

4.3.2.2 The case of an inland county

The buy-in option is found to be used extensively in a county in central China visited for fieldwork in 2017 and 2019 (referred to as County X hereafter). County X has a socioeconomic profile typical of many regions in the hinterland: relatively poor in per capita income, large outflow of working-age population, local public finance heavily relying on transfer payment, and annual deficits in local pension funds with contributions insufficient to pay for pension benefits. The county government has been under some pressure from the provincial government to meet the collection quota for pension contributions in recent years. Unwilling to tighten the collection on local firms, the county government found a solution to increase revenues by selling the entitlement to pension benefits to local hukou residents through the buy-in option.

The amount of buy-in fee depends on how many years of pension contributions the person is still short of the 15 years required. According to the interviewee, the buy-in fee for both female and male residents without any prior contribution history was 94,000 yuan, and the corresponding monthly pension benefits amounted to 1,050 yuan in 2018 in the county, with a final lump sum benefit of 42,000 yuan payable to the family upon the death of the retiree. Once the buy-in fee is made, a female resident will start to receive pension benefits when she reaches 55, while the benefits will start for a male resident when he reaches 60. Assuming that there will be no increase in the pension benefits, it will take only 4.1 years to break even from the buy-in option, i.e. for the sum of pension benefits received (including the death benefits) to equal the buy-in fee.[24] Based on the latest official

Table 4.1 Actuarial present value for the buy-in option for the UES for females and males in County X in 2018

	Buy-in fee for 15 years of pension contributions	Retirement age	Monthly pension benefits	Death benefits	Break-even age	Break-even probability	Actuarial present value
Females	94,000	55	1,050	42,000	59.1	98.9%	631,655
Males	94,000	60	1,050	42,000	64.1	96.6%	436,480

Sources: Fieldwork interview in County X in July 2019 and my calculation.[25]

data on mortality rates in China, the probability for break-even is 98.9% for females and 96.6% for males. The actuarial present value for the buy-in option is calculated to be 631,655 yuan for females and 436,480 yuan for males. The expected returns for the buy-in are 6.7 times and 4.6 times as large as the buy-in fee for females and males, respectively. With the numbers hugely in favour of those who can exercise the buy-in option, it is unsurprising that many ordinary residents intuited the financial significance of the deal and were eager to join the UES this way had they not been covered already by the pension system. Some key figures related to the buy-in option are summarised in Table 4.1.

It is obvious from the calculation that the buy-in policy, as priced and structured currently, will cost the pension system dearly in the future. Asked about whether the government across different levels had made any preparation for coping with the hidden future liability incurred by implementing the buy-in policy, the interviewee said that nothing specific was done so far to his knowledge; no funds were set aside and no transfer payments were earmarked for the purpose of meeting such future liability. In the short term, however, allowing the residents to buy in can substantially increase the revenues for the UES in the county, thus alleviating the current financial strain on the local pension funds.

4.3.2.3 The case of a costal prefecture

For an entirely different reason, the buy-in option is also used comprehensively in an affluent prefectural-level city of a coastal province visited for fieldwork in 2017 and 2019 (referred to as City Y hereafter). City Y has a vibrant economy and a young workforce with many migrant workers. Sizable surpluses have been accumulated over the years in the local pension funds. When the goal of achieving a unified pension pool at the national level for the UES was mentioned in the Social Insurance Law in 2010, the local government of City Y started to greatly widen the coverage of the UES to include as many local hukou residents as possible regardless of their employment status, which effectively locks in a large portion of the local pension fund surpluses to serve local interests in the future. Those with agricultural hukou and those with urban hukou but without formal employment, who were the target participants for the BRS as planned by the central government, were encouraged to join the UES. The local hukou residents that were not yet covered by the UES were subsidised for the pension contributions or buy-in fees required to join the UES.

Such subsidies were provided by the prefectural- and county-level governments in City Y without asking for financial help from higher-level governments at the provincial or central government levels. These subsidies were not obtained from using the existing surpluses of the local pension funds. The prefectural government gave different levels of subsidies to some 20 counties or districts of City Y, with some rich counties or districts receiving zero subsidies from the prefectural government. Besides the county-level governments, some rich villages provided substantial amount of subsidies to their residents. Besides the financial aid from subsidies, bank loans were also arranged for those approaching the retirement age to pay the buy-in fees, with the future pension benefits used as collateral for the loans.

The prefectural government also took other measures besides the subsidies and bank loans to make the pension contributions and the buy-in fees for the UES more affordable for the local hukou residents. The prefecture allowed a lower than usual lower limit for the permissible salary range. The interviewee (a director in the bureau of human resources and social security in a county in City Y) was not willing to disclose the exact percentage, but he said that it was far lower than the usual level of 60% of the local average wage. He quoted the slogan then used to describe the gist of the local policy: "lower barrier, wider coverage". He mentioned that City Y was once named as an alarming example during a work conference on pension policies held by the provincial government and reprimanded for having a particularly low level of the lower limit for the permissible salary range. Besides the financial help from the local government, the procedural requirements for joining the local UES were cut to the minimum for residents with local hukou; they were automatically eligible for buy-in when approaching the retirement age. This was in contrast to the eligibility policies implemented in most other regions such as County X, which required the residents to provide some form of proof for former employment history to be eligible for buy-in.

The implementation of such policy resulted in an extremely high level of enrolment of the UES in City Y. The total number of people covered under the UES was 2.29 million in City Y in 2018, which was very high compared to both the population with local hukou at 1.77 million and the enrolment size of the local BRS at only 221 thousand. According to the interviewee, many local rural residents who worked as farmers for their entire lives and were never employed by any companies had been enrolled into the UES and able to receive over 1,000 yuan of pension benefits per month. He also claimed that more than 97% of the retirement age population with local hukou had been covered by the UES, which is supported by the statistics released by the local government that, among the 339 thousand people receiving pension benefits in City Y in 2018, only 1.3 thousand people were receiving the benefits from the BRS. In other words, excluding those covered by the public sector pension scheme, 99.6% of people receiving pension benefits in City Y were covered by the UES. Such ratio was only 35.9% for the province in which City Y is situated and 42.5% for the country. What City Y had essentially achieved was upgrading the local BRS into the local UES. With the number of beneficiaries of the BRS now vanishingly small, there were virtually only two pension schemes left in City Y, namely the UES and the public sector scheme.

The proportion of retirement age population receiving pension benefits from the UES in City Y was even higher than that in Shanghai, which was more advanced than City Y in terms of socioeconomic development and urbanisation process.[26] Such mismatch is largely due to the differences in pension policies adopted by the respective local governments. Shanghai implemented its own version of upgrade to the local BRS. Instead of trying hard to convert every resident with local hukou into participants of the UES, Shanghai focused on improving the benefits level of its BRS.

The average monthly benefits for the BRS in Shanghai reached 1,126 yuan in 2018 (Bureau of Human Resources and Social Security of Shanghai 2019), which was 7.4 times as high as the national average at 152 yuan (Ministry of Human Resources and Social Security 2019). In City Y, the average monthly benefits for the BRS was 149 yuan, which was even lower than the national average. But most residents with local hukou in City Y could enjoy comparable, if not even higher, level of pension benefits compared to the beneficiaries of the BRS in Shanghai, because the absolute majority of them were receiving benefits from the UES in City Y now. In Shanghai, the improvement of the benefits of the BRS was predominantly financed by government subsidies, as 6.75 billion yuan, or 89.6%, of the 7.53 billion yuan of the total revenues of the BRS in 2018 came from the subsidies. The municipal government of Shanghai provided most of the total subsidies for the local BRS, as only 270 million yuan, or 3.9%, of these subsidies was from the central government (Bureau of Human Resources and Social Security of Shanghai 2019).

Both City Y and Shanghai have mainly relied on their own public financial capacity to make some material changes to the landscape of their local pension system, which has greatly benefited their local hukou residents. But unlike Shanghai injecting substantial amount of subsidies directly into the BRS to beef up its benefits level, the approach adopted by City Y to get almost everyone onboard with the UES will generate a long-lasting financial impact concerning the level of future claims by local populace on the pension fund surpluses.

While the prefectural government of City Y did not use surpluses of local pension funds to subsidise those outside of the UES for paying the required pension contributions or buy-in fees to join the UES, such policy of lowering the barrier to widen the coverage of UES will effectively reduce the amount of surpluses of local pension funds in the years to come, as the increased number of UES participants will increase the expenditures for pension benefits in the future. This is in line with the perverse incentive structure that causes the local governments to be unwilling to let the pension surpluses keep growing once they have reached a certain level, as the local governments anticipate that these surpluses will be handed over to higher-level governments sometime in the future.

The implementation of such policy in City Y serves to lock in a substantial part of the accumulated surpluses in the local pension funds by increasing the future liabilities of the local pension funds, while also benefiting the local hukou residents and thus improving the social harmony in the prefecture. The increase in pension benefits for the local hukou residents resulting from such manoeuvre by the prefectural government is also conducive to boosting domestic consumption in the prefecture and thus in line with the motive of the local government to stimulate the growth of the local economy.

Based on the valuation of the buy-in option calculated in the case of County X, the average actuarial present value for the total benefits from UES for both females and males was 534 thousand yuan in 2018. Had City Y not adopted its policy to maximise the coverage of the UES, the ratio of its retirees receiving benefits from the UES should be similar to that for Shanghai, i.e. at about 89.7% rather than 99.6% (see Endnote 26. The number of extra retirees of the UES in City Y resulted from its policy to push for wider coverage by the UES can then be estimated to be 33.6 thousand in 2018. As each of these retirees is expected to cost 534 thousand yuan for the total benefits paid throughout their retirement lives, such policy already produced a hidden future liability equivalent to 17.9 billion yuan in 2018, or 38.3% of the accumulated surpluses for the UES in City Y. Assuming that the local governments in City Y paid 50% of the required pension contributions or buy-in fees for these 33.6 thousand local hukou residents to join the UES, it would have only costed the local governments a total of 1.58 billion yuan. Considering that the extra pension benefits it brought about were worth 17.9 billion yuan, investing in such policy generated a return of 11.3 times the original investment!

City Y leveraged on the robustness of its local public finance to subsidise the relatively poor individuals with local hukou to join the UES, and it paid for such subsidies out of the public coffer of the local governments at and below the prefectural level. In combination with other clever manoeuvres such as lowering the lower limit of permissible salary range, minimising or removing the requirements for the eligibility for buy-in and arranging for government-backed bank loans for paying buy-in fees, City Y avoided the apparent misconduct of using the existing pension fund surpluses inappropriately, while solidifying its claim for the future use of a significant portion of the surpluses. The huge return of such pension policy gambit was the prize for a game well played in the convoluted arena of public finance across levels of government in China, at the cost of the long-term financial health of the pension system.

4.4 Inefficiencies caused by fragmentation

With highly fragmented pension financing and over 2,000 separately managed pension pools around the country,[27] the level of risk-pooling is low compared to what may be achieved with a nationally unified pension system. This low level of risk-pooling not only defies the principle of the law of large numbers, which underpins the financial viability of all insurance policies, but also leads to unnecessarily high administrative overheads. Empirical studies have confirmed that increases in pension plan size lead to decreases in costs per participant; therefore, centralisation can help save costs (Ghilarducci and Terry 1999). For example, the work involved in pension fund deficit settlement alone adds a nontrivial quantity of administrative costs (in terms of manhours spent in the negotiation process between levels of governments), which cannot be streamlined because it is often performed in an ad hoc manner on a case-by-case basis (as discussed in Section 4.2). The potential to achieve reasonable investment returns is also handicapped by the division of pension funds into smaller pools, because the pension fund reserves of each pool are managed by the respective local government. In most cases, the local government was only allowed to store the reserves as bank savings or use them to buy bonds issued by the central government, generating a return that barely beats inflation.[28]

A direct consequence of fragmented pension administration is a "bloated" bureaucracy, which decreases the system's efficiency by inevitably incurring high administrative costs. Beginning at the county level, local governments at different levels have their own bureaucratic units who govern the pension pools in their jurisdictions. In 2015, there were 7,915 social insurance administrative branches at county level and above in China, who in total employed a staff of 184,469—5,451 of those branches, manned by 135,453 government employees, dealt with pensions alone (Ministry of Human Resources and Social Security 2016b). Compared to the 262 million working-age people covered under the UES in China in 2015 (Ministry of Human Resources and Social Security 2016a), the US Social Security system covered 169 million workers, comprising a much smaller staff of 65,873 administrative employees and only 1,245 field offices (Social Security Administration 2017).

In each bureau of human resources and social security at the county or prefectural levels, it is common for the division that deals with the pension system to be subdivided into three arms. Each arm is in charge of one of the three pension schemes (i.e. the UES, BRS and PES) and has its own director, deputy directors and other staff.[29] As reported in the *Finance Yearbook of China 2015*, administrative costs accounted for 3.8% and 3.5% of the total expenditures by the UES and the BRS, respectively (Ministry of Finance 2015). By contrast, the administrative costs for the US Social Security pension fund were only 0.4% of the total expenditures in 2015; since 1985, the ratio of administrative costs to total expenditures for the US pension fund has remained below 1% (Social Security Administration 2019). In some regions, where the collection of pension contributions was managed by tax offices, a handling fee was charged by the tax offices, which typically ranged from 0.5 to 1% of the total amount collected.[30] This charge alone is already comparable, in proportion, to the total administrative costs of the US Social Security system.

The system does not only suffer from inefficiencies due to a "bloated" bureaucracy. In some cases, it is blighted by outright corruption (e.g. scams) that leads to huge wastage and loss of funds. An article by Xinhua News Agency states that, according to incomplete statistics from 1998 to 2005, the total quantity of social insurance funds recovered from misappropriation across the country amounted to 16 billion yuan (Xinhua News Agency 2006). In a high-profile 2006 Shanghai case, 3.2 billion yuan of local social insurance funds were misappropriated and invested in a private company (Legal Daily 2006). From 2005 to 2015, the deputy director of a district social security bureau in Zaozhuang city in Shandong embezzled 36.5 million yuan of payments for pension contributions and buy-in fees from 937 people (Xinhua Net 2017a). In 2015, an employee of the social security bureau in Yangchun, a county-level city in Guangdong, was caught soliciting and receiving bribery totalling 2.9 million yuan; he sold the entitlements for pension benefits to 61 individuals who were ineligible for such entitlements

(People's Daily 2015). From 2010 to 2017, a director in the division in charge of the disbursement of pension benefits in the Department of Human Resources and Social Security of Hebei manipulated the records in the pension information system, channelled 72.5 million yuan of benefits into his personal accounts and spent much of the stolen funds (The Beijing News 2019). In the Luohe prefecture of Henan, nine individuals, who were apparently unaffiliated with the local social security bureau, stole more than 40 million yuan of buy-in fees from over 900 local residents by claiming that they had connections with officials in the social security bureau who were able to enrol these people into the UES. The suspects went through an entire fabricated "application procedure" with the victims, including filling out forms, taking photographs, issuing social insurance cards and even accompanying the victims to the social security bureau for the collection of fingerprints as proof of personal identity (Xinhua Net 2018b).

In a 2016 incident in Hebei province, an employee of the social security bureau of the prefecture of Chengde embezzled over nine million yuan of pension funds. The employee misappropriated the lump sum cash payments received from many unsuspecting local residents as buy-in fees for the UES. To the consternation of many, including the victims, the 248 defrauded retirees were asked by the social security bureau to compensate for this loss to have their pension benefits continued. Another employee of the social security bureau of a county in Chengde committed similar crimes, stealing approximately two million yuan of pension funds from 2009 to 2013 (Jianpeng Xiong and Wang 2016). The fact that civil servants working for local social security bureaus are bold enough to take advantage of governance loopholes in the system is indicative of the lack of checks on local government conduct in general. It demonstrates that delegating the provision of pension services to local governments is riddled with inefficiencies and prone to misconduct. Further, local governments face a dilemma when such frauds are exposed: they either choose to take the responsibility and bear the costs to cover for potentially massive financial losses or request that victims pay the shortfalls (as in the above-mentioned Hebei incident). The latter option is bound to erode public trust in both the pension system and the government more broadly. In either case, the financial health of the pension system is damaged, at least temporarily. Greater harm is likely generated in the latter case because the loss of public trust will cast a long shadow over public willingness to make pension contributions.

The pension contributions required by the Social Insurance Law, which used to be 28% of wages and were reduced to 24% in 2019, are very high and cause significant financial burdens on those companies in full compliance with the law.[31] Although the ageing population is a main contributing factor for such a high rate, the inefficiencies resulting from a highly fragmented pension system are also partly to blame for the excessive costs of full compliance. If all employees participating in the UES make contributions at the same rate, with an employee-to-retiree ratio of 2.65 (in 2017), the contribution rate can be as low as 22.5% of wages. This would realise a pension replacement ratio of 60% and save the government 800.4 billion yuan of subsidies in that year, which could be used to further reduce the contribution rate to approximately 17% and therefore ease the burden on both employees and employers.[32]

Social insurance fees, including pension contributions, have become a major component of public financial system revenues in China. At 5.8 trillion yuan in 2018, the total revenues from social insurance fees were comparable in size to those from other major components of national public revenues, such as the revenues from value-added taxes (6.2 trillion yuan) and from land sales (6.3 trillion yuan). Accounting for 69% of social insurance fees, pension contributions amounted to four trillion yuan and were greater than the revenues from corporate income taxes—3.5 trillion yuan (Ministry of Finance 2019a, 2019b, 2019c). Making the pension system as cost-effective as possible would have great significance for maintaining the competitiveness of individual firms and the economy more broadly.

China has gained the title of the "factory of the world" and will likely remain a key hub for global manufacturing industries for some time. The comprehensive network of industrial clusters developed in China has benefited the economy with improved efficiency and productivity, achieved through economy of both scale and scope. This has also afforded the country a certain competitive edge in global markets. However, costs are now increasing. If operational costs in these industries increase too rapidly and, therefore, become unaligned with productivity growth, this will threaten the profitability and even viability of the industries and negate the advantages China has accumulated to date through building such cluster networks. Further, China's overall economic competitiveness will suffer. Many think that such hikes in wages are a natural consequence of China surpassing the Lewis Turning Point with diminished surplus labour. Other key reasons should also include an inefficient social security system that is financially burdensome. Under the current system in China, costs arising from pension contributions constitute a major cost component that burdens many firms, including state-owned enterprises and private companies.[33] China exhibits one of the highest pension contribution rates in the world; this clearly makes it difficult for the economy to maintain its competitiveness, which requires further maintenance and nurture for China to become a developed country. The pension contribution rate could be lowered if more working-age adults and, particularly, young migrant workers participate in the system and pay the contributions; further, a lower contribution rate could encourage more employees to join the pension system, thereby creating a virtuous circle.

4.5 Conclusion

As can be seen from the analysis in this chapter, the fragmentation in pension administration and financing has generated significant negative effects on the sustainability of the pension system in China. Such detrimental effects went beyond the inefficiencies created by the fragmented system, which nonetheless are by themselves seriously damaging the financial health of the system. Since the major pension reform launched in 1997, local governments as the provider of pension services and the manager of local pension funds have largely failed to truthfully comply with the national pension policy on the collection of pension contributions and other rules important to ensuring the financial health of the pension system. To minimise the burdens on the local economy as much as possible when collecting pension contributions is a shared characteristic for local governments across the country, as the local political leaders have the incentive to promote faster growth and can effectively control the functional units of local bureaucracies in charge of the pension system. The lack of collection efforts by the local governments results in widespread evasion and underpayment of pension contributions by firms and employees, substantially weakening the self-financing capacity of the pension system.

A more insidious threat is posed by the silent inflation of future pension liabilities caused by various kinds of manipulations on the pension pools by the local governments, such as making it easier for employees to retire and increasing the benefits level whenever expedient. As shown in the case studies, the local governments of both a poor county with pension deficits and a rich prefecture with pension surpluses have been making extensive use of the buy-in option to let more local hukou residents join the UES and receive the pension benefits. Although out of different motives, these behaviours of the local governments all contribute to the increase in the future liabilities of the pension system beyond the level that is caused by population ageing alone. The cash-based accounting method used for reporting social insurance funds asks no question about the value of future liabilities of the pension funds, which makes these behaviours of the local governments even more dangerous than the outright noncompliance to pension contribution rules, as the long-term financial impact of these local manipulations is effectively hidden from public oversight. Without a more accurate appraisal of the financial standing of the pension funds based on accrual accounting and actuarial valuation, it is possible that the higher-level governments even including the central government may not have fully grasped the consequence of such manoeuvres by their lower-level counterparts.

These self-serving behaviours of the local governments to engage in the gaming of the system are not isolated events happening only under exceptional circumstances. Rather, they are the inevitable results of a fragmented pension system with disarticulated intergovernmental relations in responsibility sharing in pension finance. And they tend to spread to other regions where the local leaders there find it politically profitable to adopt similar measures. Already, the practice of selling eligibility for pension benefits through buy-in options was found in two other counties visited during the fieldwork in the same province where the first case study is based. Several neighbouring prefectures of the city in the second case study seem to have made similar efforts to maximise the coverage of the local UES to lock in their equally sizable pension fund surpluses, as indicated by the disproportionately high ratios of retirement age population receiving pension benefits from the UES there.[34] A more concrete example of how far such "policy diffusion" can reach is the widespread practice of allowing firms to pay pension contributions according to the imputed wage at the lower limit of the permissible salary range, a unison of virtually all local governments in collective defiance of the central government policy rules on pension contributions.

With the fragmented pension administration and financing, the long-term financial sustainability of the pension system depends on the soundness of management in each one of the over 2,000 pension pools mostly in the hands of county-level governments. Yet, the existing incentive structure for the local governments as the provider of pension services and the manager of the pension funds in their jurisdictions is ridden with problems caused by moral hazard and therefore does not nurture good behaviours but rather systematically reward bad ones. And the behaviours that damage pension sustainability often get multiplied and propagated. It only takes one local leader to find a bold and innovative way of interpreting the central government policy, such as using the local average wage that was three years old to calculate pension contributions while using the most updated local average wage to calculate benefits, or upgrading the BRS into the UES via subsidised buy-ins. Other local governments wait and watch. If such new ways of taking advantage of the loopholes in the system benefit the inventor without incurring punishment from higher-level governments, they will be copied by other local governments. The detrimental aggregate effect of local implementation of pension policy on pension sustainability is thus caused by the moral hazard arising from the principal-agent problem stemming from the fragmentation in pension administration and financing, reinforced by a positive feedback loop compelled by the race to the bottom.

Notes

1 For example, the enrolment size for the UES in Jixi county of Anhui was 27 thousand as of August 2019 (People's Government of Jixi County 2019).
2 It is common for a prefectural government to issue local policy documents on how pension policy should be implemented by county-level governments in its jurisdiction. For example, a search performed in December 2019 on the portal for publicly available policy documents issued by the government of Foshan, a prefecture in Guangdong, returned 11 local pension policy documents issued from 1998 to 2018, available at https://www.foshan.gov.cn/zwgk/zcwj/. A search performed in December 2019 on the website of the Bureau of Human Resources and Social Security of Changde (http:// mohrss.changde.gov.cn), a prefecture in Hunan, also returned numerous local pension policy documents issued by the government of Changde to regulate the implementation of pension policy in the counties in Changde.
3 The Social Insurance Law promulgated in 2010 contains only 720 words (less than 1,200 Chinese characters) by transliteration on pension insurance, as compared to the analogous law on pension reform in Russia in 2013, which contains almost 20,000 words with detailed formula for pension benefits to spell out the pension policy rules universally applicable across Russia (Remington 2018). The counterpart legal documents on the social security system in the US, the UK and Australia contain 103 thousand, 23 thousand and 185 thousand words, respectively. The legal documents counted are Social Security Act (Amendments of 1965) for the US, Social Security Act (1998) for the UK, and Social Security Act (1991) for Australia.
4 Interview with an expert on Chinese pension reform in Beijing (July 2017).
5 For example, the government of Hunan issued the "Administrative Measures for the Budget of Basic Old-Age Insurance Fund for Enterprise Employees in Hunan Province" in 2019, which states that the governments at the provincial, prefectural and county levels are the main bodies responsible for the collection of pension contributions for enterprise employees and the payment of pension benefits for enterprise retirees, and that the governments at the provincial, prefectural and county levels shall undertake the task of providing financial subsidies as stipulated in the "Measures of Hunan Province for the Sharing of Responsibilities for the Basic Old-Age Insurance for Enterprise Employees" (People's Government of Hunan 2019).

6 Interview with a local official in Shandong (August 2017).
7 At the start, the percentage equalled to 2.7%, as the formula for the contribution to the CAF contains a discount rate of 90% to the provincial average wage and a multiplication ratio that was initially set at 3% and to be increased gradually in the coming years. The ratio was adjusted to 3.5% in 2019, 4% in 2020 and 4.5% in 2021 (State Council 2019; Ministry of Human Resources and Social Security, Ministry of Finance, and State Taxation Administration 2020; Ministry of Human Resources and Social Security and Ministry of Finance 2021).
8 Before the tax sharing system reform was implemented in 1994, the central government had little information about local tax bases and tax effort, and resorted to fiscal contracting to ensure it could extract at least some fixed quotas of revenues from local governments (Wong 1992).
9 Out of the 21 counties or cities visited in two rounds of fieldwork in 2017 and 2019, 12 of them had deficits in local pension funds and requested subsidies from high-level governments. Local officials interviewed in these 12 counties or cities confirmed the ad hoc, case-by-case nature of the negotiation process for securing the subsidies each year to cover for the deficits.
10 The contributions for the UES consist of two components: employer contributions, which are paid into the social pooling account, and employee contributions, which are paid into the employee's individual account. See Chapter 2 for more details.
11 Interviews with local officials in charge of the pension system in Zhejiang (September 2017), Sichuan (September 2017), Hunan (October 2017) and Guangdong (July 2019).
12 A top executive in a state-owned enterprise is often a party cadre who has a bureaucratic rank and a political career to aspire to. To them, it is more important to obey these apparent rules than to try to increase the profit margin by a few percentage points by obviously violating the Social Insurance Law. As remarked by a director of a state-owned enterprise interviewed in Hunan (July 2019), "state-owned enterprises belong to the state and it is a basic duty for them to pay taxes and fees according to the law."
13 Interview with an entrepreneur in Shanghai (July 2017).
14 The unwillingness of local governments to clamp down hard on private firms for their pension contribution obligations is also reinforced by the fact that these small private companies help create most of the job opportunities and thus contribute to growth and social stability in their jurisdictions. The local official interviewed in the first case study in Section 4.3.2 echoed such concerns.
15 Interviews with local officials in charge of the pension system in Shandong (August 2017) and Hunan (October 2017). Special schemes were also created by local governments for industrial zones and migrant workers (B. Zheng 2012).
16 According to the regulations in force in 2015, for the UES, employers should pay around 20% of the payroll, all into the pooled account, while employees should pay 8% of their salaries, all into their individual accounts (State Council 2000).
17 Interview with a local official in Shandong (August 2017).
18 Interview with a local official in a central province (July 2019).
19 Interview with a local official in Hunan (July 2019).
20 Interviews with local officials in Shandong (August 2017) and Zhejiang (September 2017). For an example of how the phrase "welfare championship (minsheng jinbiaosai)" was used with negative connotation to warn against some local governments engaging in cross-regional comparisons in welfare policies, see the summary of a speech made by Liu Wanghong, the vice dean of Jiangsu Provincial Academy of Social Sciences, available at http://www.js-skl.org.cn/academic_conference/6515.html.
21 See, for example, the "Circular of the State Council on the Approval and Transmission of the Reform Plan by the Ministry of Finance for the Accrual Government Comprehensive Financial Reporting System" issued in 2014 (State Council 2014a).
22 See, for example, the "Notice on Further Regulating the Policy on Payment for Participating in the Basic Pension Insurance for Enterprise Employees" by Guangdong (Department of Human Resources and Social Security of Guangdong and Department

of Finance of Guangdong 2018), and the "Notice on Properly Handling Issues Concerning the Payment of Basic Pension Insurance Fees for Enterprise Employees" by Hunan (Department of Human Resources and Social Security of Hunan and Department of Finance of Hunan 2017).

23 Interviews with local officials in charge of the pension system in Heilongjiang (July 2017), Hebei (August 2017), Sichuan (September 2017), Hunan (October 2017) and Guangdong (July 2019).

24 The break-even time is calculated by taking the difference between the buy-in fee and the death benefits and then dividing it by the annual pension benefits, i.e. $(94,000 - 42,000)/(1,050 \times 12) = 4.1$ years, which is a conservative estimate, as an even shorter break-even time will be derived if any increase in pension benefits is factored in.

25 The calculation method for the actuarial present values of the buy-in option is described in Appendix B.

26 In Shanghai in 2018, there were 4.49 million retirees under the UES, while 513 thousand people received pension benefits from the BRS. Thus, the ratio in Shanghai was 89.7%, as compared to 99.6% in City Y.

27 Outside the pension pools run by each county or city, some state-owned enterprises in monopolistic positions with huge profits are still taking care of their retirees by themselves with their own enterprise-based pension funds, as seen in the case of a state-owned enterprise in Hunan (July 2017). According to a report by the National Audit Office, six state-owned enterprises were found to be still managing their own pension funds in 2011, while the employees in the construction industry in 17 provinces were still not covered by the UES but instead participating in the labour insurance organised by the local bureaus of housing and urban–rural development, according to the policies released by the then Ministry of Construction in the 1980s and 1990s (National Audit Office 2012).

28 Interviews with an expert on the Chinese pension system in Beijing (July 2017) and a manager of a district branch of a state-owned bank in Hunan (July 2019).

29 Interviews with local officials in Shandong (August 2017) and Sichuan (September 2017) and with three officials in charge of the pension system in a prefecture in Fujian (September 2017).

30 Interviews with a county official in Hunan (October 2017) and a local official in Guangdong (July 2019).

31 As documented in Section 4.3.1, many companies underpay pension contributions in practice to avoid exorbitant financial burdens; noncompliance by local governments in implementing the contribution rules set by the central government is widespread. The current official standard pension contribution rate of 24% of wages consists of 16% payable by employers and 8% payable by employees. Under the current rules set by the central government, some variations in the contribution rate for employers' contributions are allowed across provinces (State Council 2005, 2019). See Endnote 7, Chapter 2 for further details regarding the contribution rules.

32 In 2017, the number of working-age UES participants was 292.7 million, while the number of UES retirees was 110.3 million (Ministry of Human Resources and Social Security 2018), resulting in an employee-to-retiree ratio of 2.65.

33 A local official interviewed in one Hunan county (October 2017) claimed that approximately 80% of the firms in the county would go bankrupt if they were subjected to tight enforcement of contribution rules for social insurance fees; an official interviewed in another Hunan county (July 2019) estimated that strict enforcement of pension contribution rules would drive one third of local companies out of business.

34 The proportion of retirement age population receiving pension benefits from the UES in the neighbouring prefectures of City Y can be calculated using the statistics released by the relevant local governments on their websites.

5 Migrant workers and pension sustainability in China

The fragmented structure of the pension system in China has significant implications for various aspects of the system's performance, such as fairness and efficiency, which can directly or indirectly affect pension sustainability. This chapter examines the unfair treatment of migrant workers by the fragmented pension system, which in turn affects the system's sustainability. Section 5.1 documents the various forms of unfair treatment of migrant workers under the fragmented pension system and discusses the resulting implications for pension sustainability. Section 5.2 concludes the chapter by examining the overall effects of such unfair treatment of migrant workers caused by the fragmentation on labour mobility, inclusive urbanisation as well as pension sustainability, while also offering a brief discussion on policy implications.

5.1 Unfair treatment of migrant workers

Several features of the Chinese pension system compromise the system's fairness and place many migrant workers in a disadvantaged position. First, the UES contribution rules are regressive, with a minimum permissible salary level for calculating pension contributions; this forces most migrant workers to pay contribution rates that are higher than standard rates if they want to join the UES. Second, the accrued benefits are not fully portable when a UES participant transfers their pension contribution history across pension pools; this particularly affects migrant workers as they tend to enjoy lower job security and must move across regions more frequently to find employment opportunities. Third, the benefit level of the UES differs considerably across regions; again, this disadvantages migrant workers because most of them can only retire from the pension pools of their hometowns, which have much lower benefits. Fragmentation in pension administration and financing has partially contributed to the formation and persistence of the regressive contribution rules, while the limited portability and regional disparities in pension benefits are direct consequences of the fragmentation. These features of the UES discourage migrant workers from participating in the system, which affects pension sustainability by limiting the scope for widening the sources of revenues from pension contributions. The difficulties encountered by migrant workers wishing to join the UES present a challenge for achieving inclusive and sustainable urbanisation; this, in turn, hinders further development and

DOI: 10.4324/9781003182696-5

productivity growth in China. The pension system also runs the risk of adverse selection, which arises from the consequences of an unfair system. The following sections will explore issues related to the unfair treatment of migrant workers by the fragmented pension system and the implications of these for pension sustainability.

5.1.1 Regressive contribution rules

Of 288.4 million migrant workers in China in 2018, 54.8% were concentrated in the eastern coastal provinces; 65.2% were males; 50.5% worked in the service industries; 27.9% worked in manufacturing firms and 18.6% worked in the construction sector. Although the percentage of migrant workers with tertiary qualifications increased from 5.7% in 2012 to 10.9% in 2018, the majority are still relatively undereducated, with 72.5% having received nine or fewer years of schooling by 2018 (National Bureau of Statistics 2013, 2019b). As a vulnerable group that sits at the bottom of the socioeconomic pecking order in the urban areas (Zhu 2003; Pan and Chen 2009), migrant workers are particularly handicapped by a fragmented pension system with regressive contribution rules that render participation in the system unaffordable for many.

According to an official interpretation of China's Social Insurance Law,

> employees with salaries lower than 60% of the local average wage must pay contributions calculated using 60% of the local average wage as the base salary. Employees with salaries higher than 300% of the local average wage must pay contributions calculated using 300% of the local average wage as the base salary.
>
> (Ministry of Human Resources and Social Security 2012)

This permissible salary range of 60% to 300% of the local average wage was introduced in the 1990s and intended to be only applied to calculate employees' contributions into individual accounts, to prevent firms from under-reporting salary figures (West 1999).[1] Later, this range was widely used by firms as a basis to report salary figures at the minimum permissible level, to reduce employers' contributions. The practice of allowing firms to pay pension contributions according to the minimum permissible salary level—at 60% of the local average wage rather than the actual payrolls—is tantamount to giving firms and their employees a 40% reduction (on average) on pension contributions.

However, for migrant workers whose earnings are commonly below the minimum permissible salary level (as demonstrated later in this chapter), the effective contribution rates imposed on them are much higher than for such firms and their employees. A local government has little incentive to follow central government policy on pension contribution rules, because obeying the rules will reduce the incentive to invest and, therefore, damage the local economy. Such a dysfunctional incentive structure results in widespread noncompliance by local governments to allow local firms to pay pension contributions according to the minimum permissible salary level, creating a perverse equilibrium during the race to the bottom (see

Chapter 4 for further details). The formation and persistence of these regressive contribution rules are partly due to the fragmentation in pension administration and financing with disarticulated intergovernmental relations and limited supervision and monitoring capacity.

The practice of applying the permissible salary range also translates to higher contribution rates for low-wage workers and lower rates for high-wage workers. For example, the local average wage used for calculating pension contributions was 5,616 yuan per month in Chongqing in 2017 (Chongqing Daily 2017). The minimum permissible (60% of the local average wage) and maximum permissible (300% of the local average wage) salary levels for pension contribution were 3,370 and 16,848 yuan, respectively. For a person earning the local minimum wage of 1,500 yuan per month (Chongqing Social Security Online 2017), the total required contribution was 943.6 yuan, shared between the firm and the worker, at an effective contribution rate of 62.9%. For the worker, the personal contribution rate of 8%, applied to the minimum permissible salary of 3,370 yuan, amounted to an effective rate of 18%. Conversely, for those in senior managerial positions in state-owned enterprises in Chongqing, whose salaries were typically over 40,000 yuan per month (Sina News 2019), the effective total contribution rate was only 11.3% and the effective personal contribution rate was only 3%.

Migrant workers are mostly low-income earners, compared with their urban counterparts. The average hourly wage for migrant workers has been estimated to be approximately 50% to 60% of that for workers with local urban hukou (Lee 2012; Zhang et al. 2016). In 2018, the average monthly income for migrant workers was 3,721 yuan (National Bureau of Statistics 2019b), whereas the national average of the local average wage used to calculate pension contributions was 6,193 yuan per month—this is 66.4% higher than the average migrant worker income (National Bureau of Statistics 2019a). Deng and Li obtained a monthly wage distribution for migrant workers based on survey data from a sample of over 8,000 migrant workers in 15 cities across China, including Shanghai, Guangzhou, Luoyang, Bengbu and Chengdu (Deng and Li 2010). The findings demonstrate that the income distribution of migrant workers in China exhibits some common features of income distributions, such as left-skewness and a median income that falls below the mean income; this indicates that the earnings of more than half of all migrant workers are actually less than the average (mean) income of migrant workers.

Based on the wage distribution derived by Deng and Li (2010), the proportion of migrant workers in China in 2017 who were earning less than the local average wage (which is used for calculating pension contributions) was estimated to be 90.6%. Data on the income level of migrant workers in individual cities are rarely reported by local governments. However, using the limited available data released by official sources, the proportion of migrant workers earning less than the local average wage in large cities is estimated to be above 90%. For example, in Shanghai, the average monthly income for migrant workers was 2,897 yuan in 2014 (Bureau of Statistics of Shanghai 2017), whereas the local average wage was 5,036 yuan per month (People's Government of Shanghai 2014); the proportion of migrant workers earning less than the local average wage was estimated to be 91.7%. In 2017, the average monthly income for migrant workers in Beijing was 3,230 yuan (People's Government of Beijing 2018), whereas the local average wage was 7,706 yuan per month (Bureau of Human Resources and Social Security of Beijing 2017); the proportion of migrant workers earning less than the local average wage was estimated to be 96.9%. The majority of migrant workers earn less than the minimum permissible salary for calculating pension contributions; the proportion of migrant workers earning below the minimum permissible salary is estimated to be 66.8% across the country, 69.4% in Shanghai and 63.5% in Beijing (based on the data described above).[2]

Many migrant workers are self-employed and run micro-businesses, often in retail and catering in the cities. To join the local pension scheme, they must pay pension contributions solely from their own income. Many of these self-employed migrant workers simply cannot afford to pay pension contributions. As indicated in the above-mentioned income distribution data, it is common for migrant workers to earn less than 60% of the local average wage; this causes them to pay higher effective pension contribution rates if they join the local UES. To join the local UES, a high proportion of self-employed migrant workers find themselves in the untenable position of having to pay pension contributions as high as one third or more of their income.[3] For example, the 2017 national average monthly income for migrant workers was 3,485 yuan (National Bureau of Statistics 2018c). In Shanghai, the 2017 minimum permissible salary level was 3,902 per month (Bureau of Human Resources and Social Security of Shanghai 2017), whereas the minimum wage was 2,300 yuan per month (Xinhua News Agency 2018b). Those migrant workers in Shanghai who earned the national average income for migrant workers had to pay 22.4% of their income to join the local UES; those earning the minimum wage in Shanghai had to pay 33.9%. Having to survive with limited income and little job security, these migrant workers are priced out of the market for gaining access to pension protection.

Most of the migrant workers interviewed across the provinces expressed significant interest in policies related to the UES. All interviewees understood that joining the formal sector pension scheme would be life changing, in terms of the prospect of retiring with any degree of financial security, because the expected level of pension benefits from this would materially change their lives for the better. Unfortunately, most migrant workers are denied access to the type of pension scheme desired, due to regressive contribution rules. When such pent-up demand is met with undersupply of the right kind of public goods, the shortfall proves to be huge. In 2018, the number of people employed in China's non-agricultural sectors (excluding civil servants and public institutions employees) reached 573 million (National Bureau of Statistics 2019a); however, only 301 million workers (52.5%) were covered under the UES (Ministry of Human Resources and Social Security 2018). In 2018, 272 million people working in the manufacturing, mining, construction, transportation and other service industries were not participants in the UES.

Migrant workers constitute the majority of this disenfranchised group, because regressive pension contribution rules force them to pay higher-than-normal effective contribution rates; this often renders UES participation too costly and is the main cause of low enrolment rates among migrant workers. According to a recent national survey, only 17% of migrant workers were enrolled in the UES (National Bureau of Statistics 2015). One key element of future pension reform in China is improving the incentives to encourage relatively young migrant workers to join the UES (Wong and Yuan 2020). The failure to extend pension coverage to most migrant workers has significant negative impact on pension sustainability by limiting the potential for raising revenue for the system from pension contributions.

5.1.2 Limited portability of benefits

When UES participants migrate across cities and provinces, they face limited portability of pension benefits accrued during previous employment. This constitutes another way in which migrant workers, who often move around the country to find employment opportunities, are unfairly disadvantaged. The incomplete transferability of acquired pension benefits from UES participation has been identified as a crucial barrier impeding internal migration and labour mobility (Dorfman et al. 2013; Lu and Piggott 2015). The loss resulting from each interjurisdictional transfer is estimated to be above 10% of the basic pension benefits from the UES for a migrant worker, in most scenarios, and may be as high as 30% for a migrant worker returning to their home province from eastern coastal areas such as Beijing, Shanghai, Jiangsu and Guangdong (Zhang and Li 2018). It is common for migrant workers to be employed in several cities or provinces throughout their working lives[4]; therefore, the cumulative loss from interjurisdictional transfers required to maintain continuous participation in the UES may be very high.

Migrant workers often return to their hometowns to retire, because certain rules make it very difficult for them to obtain UES benefits in cities. Beyond the vesting period of 15 years for the total contribution history (State Council 1997), to claim the entitlements to pension benefits from the UES in a particular city, a migrant worker must make at least ten years of pension contributions to the UES in that city (Ministry of Human Resources and Social Security and Ministry of Finance 2009). The minimum local contribution history of ten years is sufficient to disqualify most migrant workers from receiving local UES benefits. Moreover, an age limit has been placed on migrant workers to qualify for pension benefits in the cities. The rule, informally dubbed the "4050" policy, prescribes that female migrant workers over the age of 40 and male migrant workers over the age of 50 are not permitted to transfer into the pension pool of a city or province they migrate to. They can make pension contributions into the new pension pool; however, only temporary accounts will be set up for them. When they reach retirement age, they are required to transfer the balances in these temporary accounts back to the pension pools from which they have transferred (Ministry of Human Resources and Social Security and Ministry of Finance 2009).

Many migrant workers who have participated in the UES will have spent a substantial portion of their most economically productive years in cities, making contributions to both the economic development of these cities and the surpluses of local pension funds.[5] However, very few can obtain pension benefits from the UES in these cities due to restrictions on benefits. According to an official interviewed in a coastal prefectural-level city, where migrant workers constitute more than half of the local workforce and approximately one third of working-age local UES participants, the proportion of migrant workers of those receiving pension benefits from the local UES was negligibly low at below 1%.[6] In an inland county visited to conduct fieldwork presented in this study, a local official said during the interview that for every yuan the county government spent, 80 cents came from transfer payments (from higher-level governments). A key form of work performed at the local bureau of human resources and social security was to accept the pension contribution history in other provinces for the returning migrant workers in their late 40s and above. The average pension benefits for the UES in this county was less than one third of that in Shenzhen, a large costal city in Guangdong, where many migrant workers had previously worked.[7]

The limited portability of pension benefits and the barriers to obtaining UES benefits in the cities, such as the minimum local contribution history and the "4050" policy, further decrease migrant workers' interest in participating in the pension system. Even those migrant workers who can afford to pay pension contributions are apprehensive about joining the local UES of the cities where they work. One interviewee, a migrant worker from Hunan in his mid-30s, received tertiary education and was able to earn more than 10,000 yuan a month as a middle-level factory manager—a salary substantially above the local average. However, he also admitted that he was not enrolled in the local UES. During 15 years of employment in three coastal provinces, he had made pension contributions for only four years. He had decided that, whenever possible, he should avoid any employment contract that involved paying pension contributions, because he preferred to receive more cash from his salaries. He also understood the difficulty in transferring his contribution history across cities and provinces.[8] Due to these unfavourable conditions, migrant workers often choose to take cash payments for wages; this matches collusion, solicitation or even coercion on their employers' parts. This has led to the loss of substantial portions of future retirement income that should have been accrued for such migrant workers; further, the evasion of pension contributions required for the UES negatively affects the financial sustainability of the entire system.[9]

A concerning trend has begun within the pension system in recent years—the ratio of contributing employees to registered participating employees in the UES has been declining, from 86.5% in 2010 to 80.3% in 2015 (Ministry of Human Resources and Social Security 2016b). The number of employees contributing financially to the system did not increase as rapidly as the number of employees enrolled in the UES; this signals a problem of increased discontinuation of contributions among participating employees, which directly affects pension sustainability by reducing revenue from contributions. This issue of increased discontinuation of contributions is particularly concerning, as it may indicate a deeper problem of lack of trust in the government or the financial viability of the pension system. However, one of the main factors contributing to such increase in discontinuation is the fragmented nature of the pension system that is deficient in portability and transparency; this adds another layer of discrimination against migrant workers who are more mobile and often need to travel across provinces to find new employment opportunities.

The fragmented pension system is poorly adapted to catering for the varied needs of migrant workers who often have to move to other regions to secure employment opportunities. As a result, it is common for those migrant workers who previously participated in the UES to discontinue pension contributions following finding new jobs across cities or provinces.[10] Pension pools are managed by local governments, with different ground rules on contributions and benefits (as discussed in Chapter 4); therefore, interjurisdictional transfers can involve complicated and time-consuming administrative processes. A policy proposal made by the Hainan Provincial Committee of Jiu San Society (a political party in China) posited streamlining the overly complicated administrative procedures for interjurisdictional transfers of pension contribution history as a measure to help solve the problem of discontinuation of pension contributions by migrant workers (Hainan Provincial Committee of Jiu San Society 2016). Data released by the Ministry of Human Resources and Social Security demonstrated that only 20% of employees in the pension system who applied for transfers across provinces succeeded in such transfers in 2011; the other 80% failed and, therefore, discontinued their contributions (Ministry of Human Resources and Social Security 2016b).

5.1.3 Regional disparities in benefits

Another consequence of the fragmented pension system is considerable regional disparities in pension benefits for the UES. Again, this places migrant workers in a disadvantaged position, because the majority of migrant workers covered by the UES are only able to receive relatively low levels of benefits from the pension pools of their rural hometowns (as discussed in the previous subsection).

The average level of pension benefits differs considerably across provinces. In 2015, Chongqing, Sichuan, Jilin, Jiangxi and Hunan had the lowest average UES benefits, which were all below 2,000 yuan per month. Conversely, Tibet, Shanghai and Beijing had the highest average benefits—approximately double those in the five provinces with the lowest benefits (see Figure 5.1).[11] Among the six provinces with the lowest benefits, five (Chongqing, Sichuan, Jiangxi, Hunan and Anhui) are also the main net-exporters of migrant workers within China (Xinhua Net 2017b).

Further, benefits levels also tend to vary significantly across regions even within the same province. For example, among the 17 prefectural-level regions of Shandong, Rizhao had the lowest average monthly UES pension benefits (1,895.6 yuan in 2016), whereas Jinan had the highest (3,155.3 yuan). In Shandong, the highest average benefits level was 66.4% higher than the lowest level, at the prefectural level (Bureau of Statistics of Shandong 2017). The average benefits level for each prefecture of Shandong in 2016 is shown in Figure 5.2.

Data on the average benefits level for the UES at the county level are not publicly available. However, the local average wage for urban employees can be used as a proxy because it is a key parameter for determining the local benefits level. As shown in Table 5.1, the local average wage can differ by a factor of two or more across counties within the same province. In 2016, the ratio of the highest to the lowest county-level local average wage was 3.3 in Guangdong (an eastern coastal

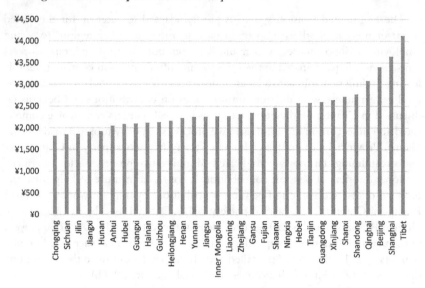

Figure 5.1 Average monthly UES pension benefits for 31 provincial-level regions of China in 2015.

Sources: China Labour Statistical Yearbook 2016 (National Bureau of Statistics 2016a).

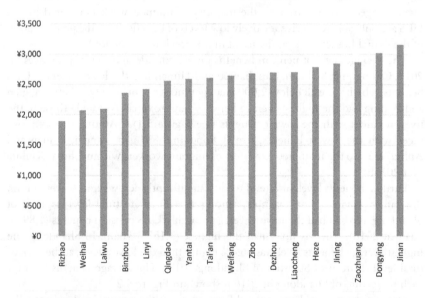

Figure 5.2 Average monthly UES pension benefits for 17 prefectural-level regions of Shandong in 2016.

Sources: Shandong Statistical Yearbook 2017 (Bureau of Statistics of Shandong 2017).

Table 5.1 Range of county-level local average wage for urban employees in Guangdong, Hunan, Sichuan and Heilongjiang in 2016

Province	Region	Lowest country-level region	Lowest local average wage (yuan)	Highest country-level region	Highest local average wage (yuan)	Highest to lowest ratio
Guangdong	Eastern	Leizhou	3,060.2	Nanshan	9,981.2	3.3
Hunan	Central	Junshan	2,594.6	Kaifu	8,713.8	3.4
Sichuan	Western	Yingjing	3,301.6	Seda	8,114.2	2.5
Heilongjiang	North-eastern	Tieli	2,393.3	Datong	6,515.8	2.7

Sources: Guangdong Statistical Yearbook 2017 (Bureau of Statistics of Guangdong 2017); Hunan Statistical Yearbook 2017 (Bureau of Statistics of Hunan 2017); Sichuan Statistical Yearbook 2017 (Bureau of Statistics of Sichuan 2017); Heilongjiang Statistical Yearbook 2017 (Bureau of Statistics of Heilongjiang 2017).

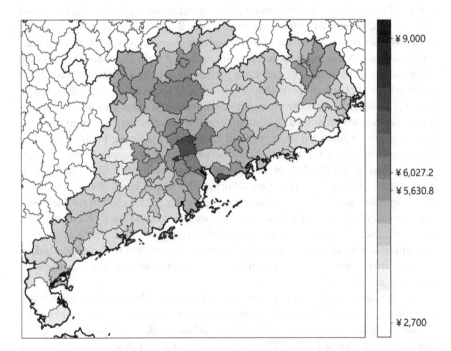

Figure 5.3 County-level average monthly wages of urban employees in Guangdong in 2016.
Sources: Guangdong Statistical Yearbook 2017 (Bureau of Statistics of Guangdong 2017).[12]

province), 3.4 in Hunan (a central inland province), 2.5 in Sichuan (a western province) and 2.7 in Heilongjiang (a north-eastern province).

To illustrate the geospatial distribution of the local average wage as a proxy for average pension benefits, a heatmap of the county-level average wage in Guangdong in 2016 was produced, (Figure 5.3). There exist substantial variations in the local average wage across counties, which leads to variations in the local average pension benefits. Remarkably, the 2016 levels of local average wage in most county-level regions in Guangdong (one of the most developed provinces of China) were below the provincial average, at 6,027.2 yuan or even below the national average at 5,630.8 yuan (as indicated by the regions in lighter shades). According to the data presented in the 2017 Guangdong Statistical Yearbook, the 2016 local average wage in 88% of the county-level regions in Guangdong was below the provincial average; in 77% of the regions, the local average wage was below the national average (Bureau of Statistics of Guangdong 2017). Thus, the geospatial distribution of the local average wage in Guangdong was highly skewed, with a few pockets of opulence mostly distributed in the Pearl River Delta (which has a high concentration of migrant workers); most other regions remained relatively underdeveloped.

5.1.4 Hindering inclusive urbanisation

The fragmented pension administration also hinders inclusive urbanisation, which is one of the goals set by the central government in its New-Type Urbanisation Plan (CPC Central Committee and State Council 2014), as local pension contribution history has been used as a barrier to limit migrant workers' access to urban public services. Many migrant workers are excluded from the UES by regressive contribution rules or discouraged from participating by the limited portability of pension benefits. Consequently, they are also denied access to urban public services, such as enrolling their children into public schools in the cities they live in, because local governments often set a continuous pension contribution history to the local UES as a precondition for those without local hukou to access public services in their jurisdictions. At the end of 2017, the percentage of people living in urban areas reached 58.5% (Xinhua News Agency 2018a), while only 42.4% of the total population in China were classified as "urban" by hukou status (Economic Daily 2018). In other words, 16.1% of the total population (223.8 million people) were urban dwellers without urban hukou.[13] Migrant workers and family members who lived with them in cities constitute the absolute majority of this group of second-class citizens, who do not receive equal treatment (compared to their local hukou counterparts) in terms of access to public services and other welfare entitlements (Pakrashi and Frijters 2017).

In 2015, as part of the ongoing hukou reform, a residence permit system was rolled out nationwide. The residence permit system was introduced to make public services in urban areas more accessible to urban residents without local urban hukou (State Council 2015b). The system functions as a "halfway house" between local urban hukou and pure migrant-worker status. Holders of residence permits for a particular city are able to access certain types of public services and levels of welfare benefits that are unavailable to migrant workers without the permit. However, the scope and degree of such entitlements remain restricted, relative to local hukou holders.

This reform took place as more migrant workers began to live with their family members in urban areas. The average urban migrant family size increased from 2.5 persons per household in 2013 to 2.61 persons in 2015, with more than half of urban migrant households consisting of three or more people. Further, 56.6% of migrant children were born in urban areas in 2014, compared to only 27.5% in 2010 (National Health and Family Planning Commission 2016). Currently, many school-age children live with their migrant worker parents in cities. Therefore, enabling their children to study at the local public schools is a major concern for migrant parents. While obtaining local urban hukou remains impossible for most migrant workers, the residence permit system seems to offer a viable alternative for gaining access to basic public services such as education.

However, there exist barriers to applying for the residence permits; these compromise the potential benefits of the system, particularly for those migrant workers most in need of improved treatment in cities. Various conditions must be met for someone to be eligible for the residence permit, depending on the relevant regulations set by local municipal governments, who also possess the discretion to determine what types of public services are available to the residence permit holders. In most cases, proof of pension contribution history to the local UES is one prerequisite for residence permit application. Typically, the pension contribution history required must be continuous and exceed a certain minimum period (e.g. 6 or 12 months)[14]; this makes the permit virtually impossible for a significant proportion of migrant workers, due to the unaffordability of pension contributions (as discussed above). Consequently, their children are excluded from local public schools and must either attend special schools run privately (with much poorer facilities and quality of education) or public schools in their hometowns, which necessitates being separated from their parents living in the cities.[15] Substandard educational opportunities for the children of migrant workers in turn contribute to the creation of social issues, such as the problem of 69 million "left-behind children" (UNICEF 2017). Such issues reinforce class segregation and stifle social mobility, the vitality of which China must nurture and improve to sustain further development and increase productivity. The unavailability of opportunities for migrant workers' children to receive better education in the cities is particularly regrettable, because it hinders China's growth potential by hamstringing the development of human capital in such a substantial portion of the younger generations.

Therefore, residence permits are only granted for those who can afford to pay pension contributions in a persistent and stable manner—those who are richer, better educated, more productive and, consequently, better able to fend for themselves. What was intended to be a progressive and inclusive reform measure has instead become regressive and exclusive. Unfortunately, because the residence permit system is set to persist at least until 2030 (according to current policy plans), this situation indicates no signs of improving in the foreseeable future (State Council 2017b).

The unaffordability of pension contributions for migrant workers exacerbates their lack of access to basic public services in the cities, because it has been exploited by local governments to create barriers around successful application for residence permits for migrant workers. These adverse conditions experienced by migrant workers have prompted some scholars to argue the existence of a new precariat in China (Swider 2015; Standing 2011). Therefore, it is unsurprising that, even following 40 years of reform and opening up, the majority of migrant workers still do not experience a sense of belonging as local residents in cities, even when many of them have spent more years living and working in urban areas than in their rural hometowns. In 2017, according to a survey conducted by the National Bureau of Statistics, only 38% of migrant workers considered themselves as local residents of the cities they lived in; in larger cities, this ratio was even lower: 25.3% and 18.7% for cities with populations of three to five million and five million and above, respectively (National Bureau of Statistics 2018c). Migrant workers form the main target population for policies promoting urbanisation; therefore, these shared sentiments do not bode well for the future of the urbanisation process.

As only approximately 17% of migrant workers are covered by the UES (National Bureau of Statistics 2015), the current level of pension protection afforded to most migrant workers is insufficient to support them to settle down permanently in urban areas, where the cost of living is considerably higher than rural hometowns. Most migrant workers are instead only covered by the BRS; therefore, their expected pension benefits are very limited. The average monthly BRS benefit was 127 yuan for the 156 million pensioners under its cover in 2017. This level of benefits was equal to 4.4% of the average monthly payouts for retirees under the UES (Ministry of Human Resources and Social Security 2018). If an elderly person over the statutory retirement age had to rely entirely on BRS pension benefits, he or she would have to endure a harsh existence; this income level falls under the poverty line (as defined by the government) by a considerable margin.[16] The lack of protection for old-age income makes it very difficult for those migrant workers from the rural hinterland, who have a minimum prospect of having their land acquired by local governments and thus receiving considerable amount of compensations, to become permanent residents of urban areas, thereby limiting the potential of the urbanisation process, which has been identified as a crucial driver of growth for the Chinese economy (Cai, Guo, and Wang 2016; Cai and Zhao 2012).

Against the backdrop of overall population ageing in China, the demographic structure of the migrant population is also changing. The average age of migrant workers in 2008 was 34 and reached 39.7 in 2017; the proportion of those aged 50 and above has also been increasing steadily in recent years, from 15.2% in 2013 to 21.3% in 2017 (National Bureau of Statistics 2013, 2018c). As the migrant worker population grows older, the demand for old-age financial security will grow more acute. Migrants from rural areas constitute the majority of the newly urbanised population; therefore, failure to provide adequate pension protection for them is incompatible with the goal of promoting inclusive and sustainable urbanisation This is particularly true when their demand for access to better quality public services and goods (including social security) will be increased due to demographic changes in the migrant population; this will also affect incentives for them to come to work and live in cities.

When the New Rural Pension was first introduced in 2009, many beneficiaries of the scheme were pleased by the sudden windfall of cash benefits, although the level of benefits for each rural elderly person remained low.[17] However, the scheme and its successor, the BRS, have not substantially improved benefits levels since this time; rather, the relative significance of benefits has decreased. The average quantity of monthly benefits was equivalent to 12.8% of per capita rural household income in 2009; in 2017, it was only 10.7%.[18] It would be unrealistic to expect policymakers to foresee that a scheme so widely welcomed by rural residents upon its launch would become obsolete so rapidly. However, without substantial changes to the BRS, the pension protection available to most migrant workers will remain insufficient for them to remain permanent residents in cities. The scheme should be recognised for its historical significance as a well-meaning transition policy rolled out, in part, to pave the way for further reform initiatives to create meaningful pension protection for migrant workers. However, it would be anachronistic to keep the scheme in its current form for any length of time.

5.1.5 Threat of adverse selection

The fragmented pension system also creates an environment in which adverse selection can occur, when the lax enforcement of pension policies gives people the opportunity to decide whether, when and where to join the UES. The first sign of adverse selection is a disproportionately high percentage of young employees who are not participating in the pension system. According to a report released by the Insurance Association of China, 20.1% of employees under 30 do not join the UES, accounting for 70.9% of the total number of employees not covered by the UES across all age groups (Insurance Association of China 2015).

The couriers and food delivery industry in China provides another example of how adverse selection can shape the behavioural patterns (regarding UES participation) of more than three million workers of a growing industry. This industry has been booming in recent years, in accordance with the rapid growth of ecommerce. It now employs over three million employees,[19] with an average wage of 6,200 yuan per month in 2018 (People's Daily 2018). However, each of the 15 employees from this industry, who were interviewed across five Chinese cities in July 2019, confirmed that they and their employers had not paid pension contributions or joined the local UES. According to their experiences, not participating in the UES is the industry standard. News articles confirm that low compliance with the Social Insurance Law is a common and widespread problem in the industry (Zhengyi Net 2019; Xinhua Net 2017c).

The majority of employees in this delivery industry are migrant workers under the age of 45, concentrated in large urban centres (CCTV 2021). These young, productive employees are more interested in earning more cash than in joining the pension system—retirement is a remote concern, if any concern at all. Many do not believe that they are disadvantaged by not participating in the pension system. When franchisees or subcontractors of large companies in this industry offer the opportunity to earn more than 5,000 yuan (in cash) per month, little persuasion is required to bring these workers on board, although no formal employment contracts or social insurance participation are included in the proposed deal.

The National Development and Reform Commission issued two policy documents on encouraging the development of the digital and Internet-based economy in 2020. The documents state that the rights of the couriers and food delivery workers, categorised as a type of "platform workers" that also include

Didi (the Chinese equivalent of Uber) drivers and other workers in the Internet-based gig economy, should be recognised, and that policies should be devised to protect their rights to participate in the social insurances (National Development and Reform Commission 2020a, 2020b). Such documents serve as a signal that the central government is aware of the importance of providing pension protection services for this growing group of workers that consist predominantly of young migrant workers. However, to date, substantive policy measures are yet to be rolled out to effect meaningful changes in this aspect of pension reform concerning the rights of migrant workers.

A clear pattern emerges—most young migrant workers (either willingly or coerced) do not participate in the UES, while migrant workers in their late 40s or 50s become increasingly interested in joining the UES. However, the only option for these older migrants is often to join the UES in their home counties, by paying the lump sum buy-in fees (discussed in detail in Chapter 4); they will often do this, when approaching retirement age, with the almost certain knowledge that they will receive back more than they pay. If the buy-in option remains open for another few decades—with the existing arrangements that allow local county officials to decide who is eligible for the buy-ins—almost anyone interested who has enough money will be able to buy the rights to receive pension benefits from the UES at the last minute. Thus, the sustainability of the pension system is compromised by a twofold problem: fewer contributions from the young workforce and greater liabilities created by the buy-ins. There exists concentration, rather than sharing, of risks, which is a typical result of adverse selection and may be disastrous for the system's financial health.

The substantial regional disparities in UES benefit levels (as described in Section 5.1.3) also lead to "arbitrage opportunities" for those seeking higher pension benefits, which results in another form of adverse selection involving potentially fraudulent business conduct. In Shandong, where the average UES benefits level is one of the highest in China (see Figure 5.1), some job agencies were found to create fictitious positions and employment history (for a fee) to allow their clients from other provinces to be accepted to a local pension pool, with substantially better pension benefits.[20] Similarly, fraudulent job agency activities—such as to create fictitious employer–employee relationships for the purpose of joining the local UES in large cities such as Beijing, Shanghai, Guangzhou and Shenzhen—are common and openly advertised online via ecommerce platforms. Through paying a service charge to these job agencies, on top of the pension contributions required by the local UES in a particular city, clients can enrol themselves into the local UES without needing to be physically located in that city. Some job agencies also provide services such as backdating pension contributions to increase the length of contribution history or even facilitating their clients without local hukou to make a lump sum payment for 15 years of contributions to the local UES of a city. This may occur through the buy-in option, so that clients reaching retirement age are immediately qualified for pension benefits; the fees required for these kinds of services can be very high, ranging from a few thousand to over 10,000 yuan.[21] The pension system's fragmented administration leaves many loopholes for these job agencies and their clients to exploit the system and enrich themselves. It is also a reasonable assumption that those people making use of such services are those who are rich enough to afford the fees and healthy enough to expect a relatively long retirement life, which again constitutes the key conditions for adverse selection to materialise.

5.2 Conclusion

The substantial increase in UES pension benefits in recent years has greatly improved the adequacy of the system, benefiting millions of retirees in China. However, this enhancement of pension benefits does not automatically lead to a fairer and more equitable distribution of pension income. A study based on recent survey data indicated that Chinese residents tend to enjoy a higher degree of happiness if they live in locations with a fairer distribution of income, after taking into account other relevant economic and demographic factors (Huang 2019). A pension system that produces a fairer distribution of benefits would, therefore, achieve higher efficiency by utilising the same quantity of funds to meet the objective of maximising citizens' happiness levels. To constitute a useful income redistribution device, the pension system should be progressive and, therefore, create a stronger sense of social justice. Low-income earners and migrant workers should be given extra protection, rather than being allowed to bear the brunt of the many defects of the current, fragmented system.

Migrant workers in China should be treated as key stakeholders in the urbanisation process, not only for the sake of social equality but also to ensure sustainable economic development. They should not be treated solely as a source of cheap labour in cities to fuel economic growth; increasingly, they are needed as permanent urban residents and consumers, so that recent large-scale investments in infrastructure will be meaningful and generate a lasting positive socioeconomic impact. Enabling these workers to join the UES will work as a fulcrum to elevate their experiences in the cities as long-term stakeholders and assist them to gain access to other types of public services; furthermore, widened UES coverage, with more migrant workers making contributions, will also serve to improve the financial sustainability of the pension system. To encourage participation and make the pension system more inclusive and accessible to migrant workers, the current policy regarding a minimum permissible salary level that is usually 60% of the local average wage should be altered. If any minimum salary level (for computing pension contributions) should exist at all, it should not be higher than the local minimum wage.

The central government made a plan in 2016 to convert 100 million migrants from rural areas into urban residents (Sina News 2016). For these newly urbanised residents, inadequate pension benefits will impede their permanent settlement in cities, affecting the progress of the strategically crucial urbanisation process.[22] Therefore, ensuring pension adequacy for those covered by the BRS could support economic restructuring and inclusive urbanisation. Ensuring the adequacy of retirement income for this vulnerable group of elderly people remains an urgent task for the government. The BRS benefits level should be substantially increased to provide more meaningful support for old-age income, for both rural and newly urbanised populations, in the short term; eventually, the BRS should be merged with the UES to ensure adequacy of pension benefits for everyone participating in the pension system. This is a crucial and necessary step for creating equal and fair access to pension services for all Chinese residents.

Ensuring adequate pension benefits for all retirees would have a significant effect on ensuring that ongoing urbanisation in China is a truly inclusive process. If new, growing urban centres are not only underpinned by the inorganic growth of steel–concrete structured buildings but also teeming with residents who are able to spend their wages and are satisfied with the public services an urban lifestyle should entail, the full potential of domestic consumption as a main driver for sustainable economic development will be realised. However, it will be a challenging task to ensure pension adequacy for everyone covered by the pension system in China, given the fragmented and decentralised mechanism through which pension services are provided to residents across the country.

In summary, the fragmented pension system, which exhibits significant differences in benefits levels between pension schemes and across regions and creates unfair treatment of migrant workers, is incompatible with the goal of enabling labour mobility across the country and ensuring inclusive urbanisation. Labour mobility and inclusive urbanisation are the two major driving forces of further development in China, which underpin not only the sustainability of the pension system but also sustainable growth of the Chinese economy. Improving the fairness of the pension system not only represents a desirable social objective, but it would also assist to set correct incentives. This will encourage more people, particularly young migrant workers, to participate in the system, lowering the incidence of discontinuation of pension contributions and avoiding situations in which adverse selection can occur, thereby improving the sustainability of the system. Fortunately, and somewhat paradoxically, the obvious lack of fairness in the existing pension system implies significant room for improvement in this area of pension reform— this may be a hitherto underexplored avenue through which effective enhancement of overall pension system performance can be achieved. The move to unify the system has already been set in motion by the central government and the information technologies required are readily available. The time is right to take a critical step by implementing a nationally unified pension system on a more equitable footing with fairness as a core value embedded in its design.

Notes

1 The local average wage used for setting the permissible salary range is determined by local governments, using wage data for the employees in the formal sector only; the income level for many migrant workers, who are employed in the informal sector, is not considered (Bureau of Statistics of Beijing 2017).
2 As described in Chapter 4, the minimum permissible salary was set at different levels for different types of companies in Beijing, with the lowest set at 40% of the local average wage in 2017 (Bureau of Human Resources and Social Security of Beijing 2017); this is the level used for the estimation of the proportion of migrant workers earning less than the minimum permissible salary in Beijing. If the normal level of 60% of the local average wage is used as the minimum permissible salary, the proportion is estimated to be even higher: 85.9%.

3 Interviews with a migrant family running a small restaurant in Beijing (July 2017), a migrant worker in Zhejiang (September 2017) and a migrant worker in Hunan (October 2017).

4 Interviews with migrant workers in Hebei (July 2017), Zhejiang (September 2017), Fujian (September 2017) and Guangdong (July 2019).

5 In an interview with the director of the bureau of human resources and social security in a coastal city (September 2017), he confirmed the effect of migrant workers on financing local pension funds. According to him, the local UES ran a deficit of approximately 10 billion yuan in 2010, which had to be financed through fiscal subsidies by the local government. In 2011, following the requirements of the then newly promulgated Social Insurance Law, the city government started to accept migrant workers' participation in the local pension system, which generated immediate and obvious effects on the financing of local pension funds. The local UES ran a surplus in 2011 and had not required fiscal subsidies until the time of interview.

6 Interview with a county official in Guangdong (September 2017).

7 Interview with a county official in Hunan (October 2017).

8 Interview with a migrant worker in Zhejiang (June 2017). Many migrant workers also tend to have a low level of trust in the local governments of the cities they work and live in. This lack of trust in urban institutions has been identified as one of the contributing factors for the unwillingness of migrant workers to participate in the UES (Wang 2016).

9 Cases of employers evading pension contributions for employees are widespread and particularly common in the construction, manufacturing and logistics industries, where labour costs are high. For example, Jinjiang is a county-level city in Fujian that is famous for shoe manufacturing. Feng Xu, a journalist, reported that approximately 90% of firms in Jinjiang did not pay social insurance fees, including pension contributions for their employees in 2018 (Feng 2018a). Another article by Feng documented that one migrant worker had worked in a company in Jinjiang for 12 years without being enrolled into the UES by his employer. The case was only exposed when the migrant worker was hospitalised following a medical emergency that occurred at work; it became clear that his employer had also not paid social insurance fees for the medical insurance, rendering the migrant worker liable for over 200,000 yuan of medical bills (Feng 2018b).

10 Interviews with migrant workers in Zhejiang (June 2017), Guangdong (September 2017) and Jiangsu (July 2019).

11 Tibet had the highest UES benefits, largely due to the special pension subsidies for those working in high altitude environments. For a retiree with at least 20 years of employment history in high altitude regions in Tibet, the monthly pension subsidies are equivalent to 15% of their average monthly pre-retirement salary (People's Government of Tibet Autonomous Region 2006, 2007).

12 The graph of geospatial distribution of county-level average wage for urban employees in Guangdong was produced using Python with the packages Matplotlib and Basemap, based on data from the 2017 Guangdong Statistical Yearbook.

13 This is based on a total population of 1.39 billion at the end of 2017 (National Bureau of Statistics 2018b).

14 See the implementation measures issued by municipal governments such as Chengdu, Hangzhou and Shenzhen (Chengdu Municipal People's Government 2019; Hangzhou Municipal People's Government 2017; Standing Committee of Shenzhen People's Congress 2014).

15 See China Youth Daily report on the poor conditions typical of special schools for the children of migrant workers (Li 2016).

16 The current official poverty line in China is 2,300 yuan of annual income per capita in 2010 constant price. In 2017, the poverty line was 3,335 yuan of annual income per capita in current price, or at 277.9 yuan per month, which was more than twice as much as the average monthly benefits for the BRS in the same year (Xinhua Net 2015; People's Government of Chongyi County 2016).

17 Interviews with four residents in a Hebei village (August 2017) and a migrant worker in Sichuan (September 2017).
18 The ratios of average benefits of the BRS to per capita rural household income were calculated according to data released by the National Bureau of Statistics and Ministry of Human Resources and Social Security (National Bureau of Statistics of China 2018; Ministry of Human Resources and Social Security 2017).
19 The total number of employees in the couriers and food delivery industry in China experienced further substantial increase amidst the COVID-19 pandemic in 2020, as more goods and services sold online to minimise physical contact drove up the demand for delivery services. In the first half of 2020, 2.95 million workers worked for a single delivery company (Meituan) alone (State Information Center 2021).
20 Interview with a local official in Shandong (August 2017).
21 Interview with a job agency employee (July 2019). In December 2019, I searched using some relevant key words on Taobao, a major e-commerce website in China; this returned hundreds of listings of job agencies offering such services as described here.
22 As mentioned in Chapter 1, ensuring inclusive urbanisation and labour mobility has been identified by Lou Jiwei, the former Finance Minister, as a necessary condition for China to sustain its growth and avoid falling into the middle-income trap (Lou 2015).

6 Discussion and conclusion

This chapter first reviews the key findings presented in the previous three chapters to provide a synthesised response to the two research questions on pension sustainability in China, which focus on the effects of population ageing and fragmented administration, respectively. The chapter also discusses the results in relation to the existing literature. The contributions and limitations of the study presented in this book are then outlined with some suggestions on future research directions. The chapter concludes the book by offering a brief reflection on the role of pension reform in the broader context of China's ongoing transformation into a modern market-based economy and drawing some policy implications based on the main findings of this study.

6.1 Key findings and discussion

6.1.1 Research question 1

How does the population ageing influence the financial sustainability of the pension system in China? Specifically,

(a) *What impact will the fertility trends of China generate on the sustainability of its pension system under the plausible reform options in retirement age?*

As shown in the results of the quantitative simulation of retirement age reform under the three fertility scenarios, there is an asymmetry in the effects of the fertility trends on the financial health of the pension system. Even a very optimistic fertility assumption cannot substantially improve the old-age dependency ratio compared to the baseline fertility assumption. But a persistent and very low level of fertility rate as assumed in the pessimistic scenario can lead to a marked increase in the old-age dependency ratio, causing considerable financial pressure on the pension system.

The optimistic fertility scenario can only lead to limited improvement in terms of the old-age dependency ratio. This shows that population ageing is bound to occur in China in the coming decades, even if the total fertility rate in China can be raised to the replacement level in only ten years as boldly assumed in the optimistic fertility scenario. Under the optimistic fertility scenario, the financial health of the pension system is only slightly improved as compared to the baseline fertility

DOI: 10.4324/9781003182696-6

scenario. This is reflected in the mild increase of the LTSR by 2.6 percentage points on average. Thus, there is only a limited potential for a favourable change in fertility level in China to lessen the financial burden caused by population ageing on the pension system.

The old-age dependency ratio reaches much higher levels under the pessimistic fertility scenario as compared to the baseline scenario, causing a substantial decrease in the LTSR of 8.5 percentage points on average, which can translate into a more than one third reduction in pension benefits for retirees as in one of the reform options tested when the retirement age remains unchanged at 57 with the financial balancing implemented in 2032. The pessimistic scenario thus delivers a warning signal for a potential funding crisis, for which the policy makers should be prepared.

As the recognition of such asymmetry of the effects of the future fertility trends in China on the financial health of the pension system is still missing in the existing literature, this study thus helps enrich the current debate by highlighting the possibility of a persistently low fertility level leading to a pension funding crisis.

(b) *How do different plausible reform options in financing method, retirement age and timing of implementation affect overall welfare outcome across generations while ensuring financial sustainability under different fertility scenarios?*

Under each retirement age reform option and in all fertility scenarios, some form of PAYG system turns out to be the optimal financing plan that dominates the fully funded system in terms of welfare gains across the generations. The timing of launching the reform to postpone the retirement age has some effect on the long-term sustainable pension benefits level. Delaying the reform by ten years lowers the LTSR in general, and such an effect is magnified under the pessimistic fertility scenario.

The fully funded system is never the best financing plan under all fertility scenarios and in all reform options. It therefore indicates that the best choice of financing methods for the pension system in China remains a PAYG system. However, the parametric details of the PAYG system ensuring welfare optimality, e.g. when to implement the financial balancing of the pension system and whether to make the system annually actuarially balanced, vary with the retirement age reform options, which suggests a feature of path dependence that should be taken into account in the policy making process for the new pension system design in China.

The timing of launching the retirement age reform has some impact on the long-term affordable pension benefits level. In general, delaying the reform by ten years (i.e. 2032 vs. 2022) decreases the LTSR. Under the baseline and optimistic fertility scenarios, the reductions in the LTSR are relatively small (reductions all below two percentage points). Under the pessimistic fertility scenario, the ten-year delay in activating the reform is more costly as manifested in the larger reductions in the LTSR (5.0 percentage points for retirement at 57, 3.7 for 60, and 1.4 for 65). Thus, it would be a cautious measure for the policy makers to closely monitor the fertility level and its trends in China before deciding when to implement the retirement age reform.

Postponing the retirement age to 65 seems to be able to mitigate some of the negative financial impact of introducing the reform at a later time, as the ten-year delay in implementation of the reform only causes relatively small reductions in the LTSR in each of the three fertility scenarios (0.3, 0.2 and 1.4 percentage points for the baseline, optimistic and pessimistic scenarios, respectively).

The results of this research thus refute the view that China should change the financing mechanism of the pension system from a PAYG system to a funded one in order to make the financing of the system more affordable (Feldstein 1998). The finding of this research that the PAYG system is still the preferred option for financing the pension system in China is derived based on the simulation results by the OLG model that realistically captures many salient features and interdependent aspects of China as a developing country, such as the relatively high growth rates for wages and productivity as well as the lack of well-developed capital markets, which are excluded from the calculation by Feldstein to arrive at his conclusion that a fully funded system is more suitable for China (Feldstein 1998).

(c) *Is there a fertility cliff under which the pension system will be in de facto bankruptcy with all plausible retirement age reform options?*

To determine whether there exists a fertility cliff under which the pension system will be in de facto bankruptcy with all plausible retirement age reform options, it is necessary to examine whether any reform option is powerful enough to counter the negative financial impact of the worst-case scenario. More specifically, it can be determined by looking at the LTSR achievable by the highest viable retirement age (i.e. 65) in the pessimistic fertility scenario and comparing the LTSR with a standard as a threshold for defining old-age pension adequacy.

Postponing retirement to 65 can achieve an LTSR of 43.5% if implemented in 2022 or 42.1% if implemented in 2032, under the pessimistic fertility scenario. The ILO has established a number of conventions on pension provision, which are agreed upon by a wide range of governments, employers and trade unions. A replacement ratio of 45% is set by Convention 128 of the ILO to "guarantee protected persons who have reached a certain age the means of a decent standard of living for the rest of their life".

The highest LTSR achievable in the pessimistic fertility scenario is 43.5%, slightly below the replacement ratio of 45% as set by the ILO as a threshold to protect the decency of life for the retirees. Therefore, the possibility of the fertility cliff for China cannot be ruled out. Such results should serve as a reminder to the government that they should pay more attention to monitoring the fertility rate in China and its future trends in a more timely, transparent and accurate manner. Should the fertility rate drop even further than its current level, the government should take more proactive measures to reverse the adverse trends in fertility in order to avoid a funding disaster on the pension system in the future.

The findings of this research enrich the current debate on the impact of population ageing on pension sustainability in China by assessing the financial sustainability of the system based on the notion of de facto bankruptcy, a more nuanced conception of how to assess the bankruptcy of a pension system than the nominal sense of bankruptcy that can always be prevented by adjusting internal system parameters such as contribution rate, benefits level and retirement age. By considering a wider range of fertility scenarios, this research also identifies the possibility of the fertility cliff under which the pension system will be in de facto bankruptcy with all plausible retirement age reform options, which is an area of analysis that is missing in the existing scholarship. With the threat of a persistently low fertility level in China, the retirement age reform alone is found to be only a necessary but not sufficient condition for preventing the de facto bankruptcy of the pension system. Such finding refutes several results in the literature that concur on the potency of postponing retirement age in fundamentally solving the funding crisis of the pension system in China (Zeng 2011; Zhongmei Yuan 2013; Z. Song et al. 2015), while aligning with several other studies that find the retirement age reform to be only a necessary ingredient for solving the fiscal problem posed by the fast ageing population (Yuan 2014; Yu and Zeng 2015; Yang and He 2016).

6.1.2 Research question 2

How does the fragmentation in pension administration and financing influence the financial sustainability of the pension system in China?

Since the economic reform started in China in the late 1970s, the Chinese pension system has remained fragmented with over 2,000 separate pension pools mainly managed by the county level governments, which constitutes a unique and salient feature of the system. Such fragmentation is found to have generated substantive negative impact on pension sustainability in China. Three mechanisms through which the fragmentation affects pension sustainability are identified: 1) the moral hazard that influences the behaviours of local governments, 2) the inefficiencies caused by the fragmented structure of the system, and 3) the unfair treatment of migrant workers that discourages their participation in the UES, hinders inclusive urbanisation and creates the conditions for adverse selection to occur.

1) Moral hazard influencing the behaviours of local governments
   The disarticulated intergovernmental fiscal responsibility in pension financing and deficit settlement in China has created a principal-agent problem, where moral hazard arises to compromise the financial sustainability of the pension system. The central government has provided large amount of subsidies to bail out the local pension pools in deficits to ensure the timely and full disbursement of pension benefits in all places. However, the intergovernmental fiscal responsibility for financing pension deficits has not been clearly defined in legal and policy documents, leading to an ad hoc and complicated negotiation process between different levels of governments on pension fund deficit settlement. The local governments effectively face a soft budget constraint on their local pension funds, as they are able to get enough subsidies from higher-level governments even when they fail to meet the collection quotas for pension contributions, with the central government being the ultimate guarantor of the solvency of the pension system and the main source of subsidies. Under

the soft budget constraint on local pension funds and an incentive structure that rewards the local political leaders for achieving fast growth of the local economy, the local governments behave in ways that incur systematic implementation bias to the detriment of the financial sustainability of the pension system. As shown in the summary of the findings below, this research not only confirms the existence of moral hazard in the Chinese pension system as identified in the existing literature (James 2002; Chen 2004; Lou 2019), but it also provides a more in-depth analysis of how the moral hazard arises under the fragmented system and the mechanism through which it affects pension sustainability in China. Moreover, this study helps probe into the seriousness of the impact of moral hazard on the sustainability of the system by attempting to quantify the net increase in liabilities resulting from the buy-in option exploited by the local governments in a county and a prefecture in the case studies based on the data collected from the fieldwork.

On the revenue management side, the local governments have largely failed to comply with the national pension policy on the collection of pension contributions since the 1997 reform to launch the UES nationwide. The local governments tend to minimise the burdens on the local firms as much as possible when collecting pension contributions, as the local political leaders are rewarded for promoting faster economic growth and can influence the behaviours of the functional units of local bureaucracies in charge of the pension system by controlling the disbursement of government funds for personnel, budgets and properties. The lack of collection efforts by the local governments has led to widespread evasion and underpayment of pension contributions by firms and employees, substantially weakening the self-financing capacity of the pension system.

On the expenditure and liability management side, the manipulations on the local pension pools by the local governments have produced a more insidious threat: the silent inflation of future pension liabilities. The local governments have been noted to allow early retirement and increase the benefits level whenever expedient. The buy-in option has been used extensively, as shown in the case studies, in both a poor county with pension deficits and a rich prefecture with pension surpluses, in order to enrol more local hukou residents into the UES to obtain pension benefits. The actuarial present value for the buy-in option was calculated to be 534,000 yuan in 2018 on average for both females and males in the county in the first case study, compared to the buy-in fee of only 94,000 yuan. And the policy adopted by the local government of the prefecture in the second case study to enrol as many local hukou residents as possible into the local UES produced a hidden future liability with an estimated actuarial present value of 17.9 billion yuan in 2018, or 38.3% of the accumulated surpluses for the local UES. These behaviours of the local governments have led to the increase in the future liabilities of the pension system beyond the level caused by population ageing alone.

On the issues of the fiduciary management and supervision on the pension funds, the cash-based accounting method used for reporting social insurance funds gives no estimation of the value of future liabilities of the pension funds, making such increase in pension liabilities even more dangerous than the outright noncompliance to pension contribution rules, as the long-term financial impact of these local manipulations is effectively hidden from public oversight. Without a more accurate appraisal of the financial standing of the pension funds based on accrual accounting and actuarial valuation, the higher-level governments including the central government may not have fully realised the implications of such manoeuvres by the lower-level governments.

Under the fragmented system, the long-term pension sustainability depends on the soundness of management of local pension pools in the hands of county-level governments. But the existing incentive structure for the local governments is ridden with problems caused by moral hazard. The behaviours that damage pension sustainability often get multiplied and propagated. It only takes one local leader to find a new way to take advantage of the loopholes of the system, such as using the local average wage of three years ago to calculate pension contributions while using the latest local average wage to calculate benefits, or upgrading the BRS into the UES through subsidised buy-ins. Other local governments wait and watch. If such gaming of the system benefits the inventor without incurring punishment, other local governments will follow suit. For example, two other counties in the province where the first case study is based were found to have adopted the practice of selling eligibility for pension benefits through buy-in options. Similar efforts to maximise the enrolment of local hukou residents into the UES to lock in the surpluses in the local pension funds were taken in several neighbouring prefectures of the city in the second case study, as indicated by the abnormally high percentage of people receiving pension benefits from the UES there. A more concrete example of how far such "policy diffusion" can reach is the widespread practice of allowing firms to pay pension contributions based on the minimum permissible salary rather than actual payrolls, a unison of virtually all local governments in collective defiance of the central government policy rules on pension contributions. The detrimental aggregate effect of local implementation of pension policy on pension sustainability is thus caused by the moral hazard arising from the principal-agent problem stemming from the fragmentation in pension administration and financing, reinforced by a positive feedback loop compelled by the race to the bottom.

2) Inefficiencies caused by the fragmented structure of the system
The highly fragmented structure of the system leads to low level of risk pooling, high administrative costs, low investment returns and difficulties to implement effective control and supervision, which make the system prone to corruption. These all contribute to the inefficiencies of the system, which have direct negative impact on pension sustainability.

With over 2,000 separately managed pension pools, the level of risk pooling is low compared to what could be achieved with a nationally unified pension system. The low level of risk pooling is against the principle of the law of large numbers that underpins the financial viability of all insurance policies. It also leads to high administrative overheads. For example, the work involved in pension fund deficit settlement alone adds a nontrivial amount of administrative costs in terms of manhours spent in the negotiation process between levels of governments, which cannot be streamlined as it is often done in an ad hoc manner on a case-by-case basis. The potential for achieving reasonably good investment returns is also handicapped by the division of pension funds into smaller pools, as the pension fund reserves in each pool are managed by the respective local government, which is only allowed in most cases to store the reserves as bank savings or use them to buy bonds issued by the central government, generating a return that barely beats inflation.

The fragmentation in pension administration leads to a bloated bureaucracy. Local governments from the county level and above have their own bureaucratic units for managing the local pension pools. The fieldwork shows that, in each bureau of human resources and social security, it is common for the division for dealing with the pension system to be subdivided into three arms. Each arm is in charge of one of the three pension schemes, i.e. the UES, BRS and PES, and each having its own director, deputy directors and other staff. The administrative costs accounted for 3% to 4% of the total expenditure for the UES and the BRS in recent years according to official statistics, much higher than the administrative costs for the US Social Security system that have remained below 1% since 1985. The fieldwork also shows that, in some regions where the collection of pension contributions was managed by tax offices, a handling fee was charged by the tax offices, which ranged from 0.5% to 1% of the total amount collected. This charge alone is already comparable in proportion to the total administrative costs of the US Social Security system.

Due to the difficulties for implementing effective control and supervision on the highly fragmented system, cases of corruption and fraud are common and can lead to huge loss of funds. For example, according to incomplete statistics from 1998 to 2005, the total amount of social insurance funds recovered from misappropriation across the country amounted to 16 billion yuan. In a high-profile case in Shanghai in 2006, 3.2 billion yuan of local social insurance funds were misappropriated and used to make investment in a private company. From 2005 to 2015, a deputy director of a district social security bureau in Zaozhuang city in Shandong embezzled 36.5 million yuan of payments for pension contributions and buy-in fees from 937 people. From 2010 to 2017, a director in the division in charge of the disbursement of pension benefits in the Department of Human Resources and Social Security of Hebei manipulated the records in the pension information system, channelled 72.5 million yuan of benefits into his personal accounts, and spent most of the stolen funds.

When such cases happen, the local governments face a dilemma: they either choose to take the responsibility and bear the costs to cover for the potentially massive financial losses, or ask the victims to pay up the shortfalls as in an incident in Hebei, which will lead to the corrosion of public trust about the pension system in particular and the government in general. In either case, the financial health of the pension system is damaged, at least temporarily. And greater harm will likely be generated in the second case as the loss of public trust may cast a long shadow over people's willingness to make pension contributions.

While the fragmentation of the system has been found to set a limit on the efficiency improvement of the system as early as the late 1990s (West 1999), this study confirms that, two decades later, the system remains inefficient as demonstrated by the low level of risk pooling, high administrative costs, low investment returns and difficulties to implement effective control and supervision that can lead to huge losses of pension funds and the corrosion of public trust in the system due to corruption and fraud, despite all the progress in information and communication technology that should have made the management of the system much more efficient.

3) Unfair treatment of migrant workers
Migrant workers are unfairly treated by the pension system with its regressive contribution rules, limited portability of accrued benefits and regional disparities in benefits level of the UES. The fragmentation in pension administration and financing has partially contributed to the formation and persistence of the regressive contribution rules, while the limited portability and regional disparities in pension benefits are direct consequences of the fragmentation. These features of the UES discourage migrant workers from participating, hence affecting pension sustainability by limiting the scope for widening the sources of revenues from pension contributions. They also present a challenge for achieving inclusive and sustainable urbanisation, which in turn hinders further development of the country. Moreover, an unfair system produces the conditions for adverse selection to occur, posing a threat to pension sustainability.

The contribution rules of the UES are regressive with a minimum permissible salary level for calculating pension contributions, which forces most migrant workers to pay at higher-than-normal contribution rates if they want to join the UES, as the majority of migrant workers earn less than the minimum permissible salary. For example, the proportion of migrant workers earning below the minimum permissible salary in recent years is estimated to be 66.8% across the country, 69.4% in Shanghai and 63.5% in Beijing based on secondary data. And the fieldwork reveals that a large proportion of self-employed migrant workers would have to pay one third or more of their income as pension contributions if they wanted to join the UES in the cities.

The regressive contribution rules make the participation of the UES too expensive for most migrant workers, which is a major cause for the low enrolment rate among migrant workers. A recent national survey shows that only 17% of migrant workers were enrolled in the UES. As an important element of future pension reform in China should entail improving the incentives for encouraging relatively young migrant workers to join the UES, the failure in extending the coverage to most migrant workers due to the regressive contribution rules will generate significant negative impact on pension sustainability by limiting the potential for raising revenues for the system from pension contributions.

The accrued benefits are not fully portable when a participant of the UES transfers the pension contribution history across pension pools, which particularly hurts migrant workers as they tend to enjoy lower job security and need to move across the regions more frequently to find employment opportunities. However, most migrant workers who have participated in the UES of the cities are effectively forced to transfer their pension contribution history back to their hometowns, as certain rules on the entitlement to pension benefits, such as the minimum local contribution history of ten years and the age limit for joining the UES of the cities (i.e. the "4050" policy), make it very difficult for them to obtain benefits from the UES of these cities. The fieldwork reveals that the proportion of migrant workers in those receiving pension benefits from the local UES was negligibly low at below 1% in a coastal prefectural-level city where migrant workers constitute more than half of the local workforce and around one third of the working-age participants of the local UES.

The lack of portability and the restrictive rules on benefits entitlement further dampen migrant workers' interest in participating in the pension system. The fieldwork reveals that it was common for migrant workers to prefer more cash payment for their salaries than participating in the UES and paying the contributions, even if their salaries were high enough to afford to pay the required contributions. The limited portability of accrued benefits and the complicated and time-consuming administrative process required for the interjurisdictional transfer of pension contribution history also contribute to the increase in discontinuation of pension contributions by migrant workers in recent years. Both the evasion and the discontinuation of pension contributions have negative impact on the financial sustainability of the system.

The benefits level of the UES differs considerably across the regions, which again works to the disadvantage of migrant workers as most of them can only retire from the pension pools of their hometowns with much lower benefits, despite the fact that some migrant workers have paid pension contributions at higher-than-normal rates into the UES of the cities. Among the six provinces with the lowest average benefits, five of them, namely Chongqing, Sichuan, Jiangxi, Hunan and Anhui, are also the main net-exporters of migrant workers within China. The benefits level can also vary substantially within provinces, with the highest average benefits often being more than double the amount of the lowest average benefits across the counties in the same province. While the regional disparities in pension benefits do not directly damage pension sustainability, they help create the situation where adverse selection can occur as discussed later.

As many migrant workers are excluded from joining the UES by the regressive contribution rules or discouraged from participating by the limited portability of pension benefits, they are consequently denied access to urban public services such as the right to send their children into public schools, because the local governments often require a continuous pension contribution history to the local UES for those without local hukou to access these public services in the cities. With only around 17% of migrant workers covered by the UES, most migrant workers are only covered by the BRS that provides very limited benefits. The average monthly benefit of the BRS was 127 yuan for the 156 million pensioners under its cover in 2017, which was equal to 4.4% of the average monthly benefits of the UES and less than half of the official poverty line. The lack of access to urban public services and the insufficient protection for old-age income make it difficult to convert these migrants into permanent urban residents, thus limiting the potential for the urbanisation process, which has been identified as an important driver of growth for the Chinese economy.

The fragmented pension system also creates an environment where adverse selection can occur, when the lax enforcement of pension policies gives people the opportunity to decide whether, when and where to join the UES. Most young migrant workers, either willingly or coerced, do not join or discontinue the contributions to the UES in the cities, while many migrant workers pay the lump sum buy-in fees to enrol themselves into the UES of their hometown counties immediately before the retirement age with the almost certain knowledge that they will get back more than what they pay. If the buy-in option remains open for another few decades with the local officials in the counties deciding who are eligible for the buy-ins, almost anyone with enough money will be able to buy in at the last minute. The sustainability of the pension system is then faced with a double whammy: less and less contributions from the young workforce with more and more liabilities created by the buy-ins. There is no sharing but concentration of risks, which is the typical result of adverse selection and a recipe for disaster for the financial health of the system.

The regional disparities in benefits of the UES also lead to "arbitrage opportunities" for those seeking higher pension benefits, which can result in another form of adverse selection. The fieldwork reveals that some job agencies created fictitious positions and employment history for a fee to let their clients from other provinces get accepted to the UES in cities in a coastal province with substantially better pension benefits. Similar activities by job agencies to create made-up employer-employee relationships for the purpose of joining the local UES in large cities such as Beijing, Shanghai, Guangzhou and Shenzhen are common and openly advertised online. The clients can enrol themselves into the local UES, even without the need of being physically in these cities. Some job agencies also provide services for backdating pension contributions or facilitating buy-ins. The fragmented administration of the pension system leaves many loopholes for these job agencies and their clients to exploit to enrich themselves. It is reasonable to believe that the people making use of such services are rich enough to afford the fees and healthy enough to have a relatively long life expectancy, which again constitutes the key conditions for adverse selection to materialise.

The results of this study thus align with the findings in the literature that the fragmented pension system leads to limited portability and regional disparities in benefits level (West 1999; Cai and Cheng 2014; Zhang and Li 2018), but this study also offers a synthesis of these features of the system, together with the regressive contribution rules, which underpin the unfair treatment of migrant workers by the pension system and contribute to their under-participation in the system, which in turn affects pension sustainability. Moreover, this study argues that the unfair system leads to the conditions for adverse selection to occur with the support of the examples of some local officials deciding who are eligible for the buy-in option and certain job agencies taking advantage of the system loopholes to help their clients get better pension benefits.

6.2 Contributions and limitations

This research studies the financial sustainability of the Chinese pension system by focusing on the impact of the two idiosyncrasies of China, namely the ongoing population ageing at unprecedented speed and scale and the highly fragmented structure of the pension system. As shown in the above discussion of the main findings, this research contributes to a new way of assessing the prospect of ensuring pension sustainability in China based on a deeper understanding of the effects of the demographic transition and the implications of the fragmentation. Specifically,

- This research enriches the current debate on the impact of population ageing on pension sustainability in China by assessing the financial sustainability of the system based on a more nuanced approach for determining pension scheme bankruptcy, which is built on the notion of de facto bankruptcy rather than nominal bankruptcy, as the latter can always be prevented by adjusting system parameters such as contribution rate, benefits level and retirement age.
- The wider range of fertility scenarios considered in the simulation analysis helps to assess the possibility of a fertility cliff, under which the pension system will be in de facto bankruptcy with all plausible retirement age reform options, which is an area of analysis that is missing in the existing scholarship.
- This research helps to fill the gap in the existing literature on pension reform in China by conducting a detailed analysis of the effects of the fragmentation on pension sustainability.
- As the limited availability of secondary data on pension administration and financing at local level may have contributed to the lack of in-depth analysis on the implications of the fragmentation in the scholarship to date, the primary data collected from the fieldwork afford a useful addition to the existing stock of empirical evidence for studying this topic. In particular, the primary data presented in the two case studies help reveal the seriousness of the damage the behaviours of the local governments can cause on the financial health of the pension system under the moral hazard embedded in the fragmented system with disarticulated intergovernmental fiscal responsibility.
- Besides the moral hazard that distorts the behaviours of the local governments, this research also identifies other mechanisms for the fragmentation to affect pension sustainability that are underexplored in the literature. Various types of inefficiencies caused by the fragmentation have direct negative impact on pension sustainability, while the unfair rules of the fragmented pension system work to the disadvantage of migrant workers and discourage their participation, thus hindering inclusive urbanisation and creating the conditions for adverse selection to occur.

• As alluded to in Chapter 1, these lessons learnt from studying China's recent pension reform can be useful input for policy makers in other countries already facing or expected to confront similar issues associated with the problem of "getting old before getting rich" in the coming decades. This research, as an effort to better understand China's pension sustainability, also indirectly contributes to resolving such pressing global issue.

As with most studies, this research has several limitations. Two major limitations of this study should be considered when interpreting the findings and can be addressed in future research. Firstly, the limited availability of data on the pension system in China in general and at local level in particular is still a major constraint to any scholarly efforts to gain a deeper understanding of how the system works on the ground in the political and institutional context of China. The fieldwork conducted for this research was an attempt to supplement the limited secondary data with the primary data collected from interviews with local officials, local residents, migrant workers, entrepreneurs and other groups of stakeholders of the pension system. The two rounds of fieldwork in 2017 and 2019 covered 12 provinces in all four regions of China, including seven eastern coastal provinces, one north-eastern province, two central provinces and two western provinces. As a focus of the fieldwork was to better understand the incentives and behaviours of local governments, the selection of fieldwork sites was largely driven by the availability of interview opportunities with local officials that were often difficult to obtain. Given the size and diversity of China, it would be ideal to have more provinces covered by the fieldwork with a more even distribution of provinces across the four geographic regions, thus helping generate more discoveries and insights from a more comprehensive coverage of the varying local conditions. However, the time and resources that could be dedicated to this research were inevitably limited, making such a plan to expand the scale of the fieldwork impractical under the current circumstances.

Secondly, the modelling technique adopted to conduct quantitative simulations of the impact of population ageing on pension sustainability abstracts from the effects of the fragmentation by assuming a nationally unified pension system. Therefore, the findings of the quantitative analysis should be read in light of this limitation: the ageing of the population with a persistently low fertility level carries the potential to make a pension system without all the defects caused by the fragmentation fall into de facto bankruptcy. The implication is then clear that the prospect is even bleaker for ensuring pension sustainability in China if the system remains a fragmented one. In this regard, the analyses on the two research questions complement each other. The negative impact of the fragmentation on pension sustainability is found to be nontrivial, thus informing the importance of incorporating such institutional feature in future research involving similar quantitative methods to simulate pension reform in China. The possibility of the fertility cliff and the danger of the de facto bankruptcy caused by the adverse demographic trends accentuate the urgency of the reform to defragment the system. Therefore, the findings of this research only mark the end of the beginning for the efforts the researcher plans to undertake, in order to gain a better understanding of the pension system in China and to facilitate the improvements required to ensure its long-term sustainability.

6.3 Conclusion

Until now, the steps taken for pension reform in China have reflected a general absence of overall strategy to build a well-functioning pension system conducive to the long-term prosperity of the country transitioning towards a more open and market-oriented economy. Such lack of vision and top-level design in pension reform was evident in times of fiscal decline, recovery and buoyancy as experienced through the recent history of Chinese public finance from the late 1970s until the present day. Not unlike other areas of the economic reform in China, the pension reform has proceeded in a piecemeal manner, with each step taken often as a reaction to the unforeseen problems encountered in reality or the side effects unexpectedly engendered by the previous round of reform. The reform carried out so far can be characterised by a series of ad hoc tinkerings on the pension system, which did over the years make some improvements in one aspect or another to alleviate certain symptoms most acutely afflicting the suboptimal system at a particular point of time. But the root cause for its multitude of pitfalls has so far been left untouched, as the policy makers did not summon enough resolve to revamp the still highly fragmented and decentralised system.

If the central government still chooses to cope with the various drawbacks and problems associated with the pension system by muddling through, there will be a more than remote likelihood of leaving some most challenging legacy issues such as those caused by the fragmentation unresolved for too long. The kick-the-can-down-the-road mentality carries a physical limit, as one can only keep kicking the resolution of the real issue into the future until the road ends. By then, it may be too late as the window of opportunity to reach a reasonably satisfactory solution of the problem can be missed. As a case in point of the central government's indecision so far to fundamentally address the underlying problem of the highly fragmented and decentralised administration of the provision of vital public services, the pension system may be on a collision course at the end of the above-mentioned metaphorical road if actions are not taken soon enough. The possibility of the fertility cliff revealed by this study represents one way how the road may end.

Three mega trends unfolding in China, namely the transition towards a market economy from a planned one, the demographic changes underpinned by a rapidly ageing population and the urbanisation at an unprecedented pace and scale, have been noted to collectively shape the socioeconomic contexts of the pension reform (Hussain 2007). The seismic shift in the economic structure of the country becoming a more market-based one dictates the supply side of those public goods and services by affecting public finance conditions and therefore resources available to the governments, while the dynamics of the demand side is driven by the demographic changes and the urbanisation process. In the case of the pension system, the pressure will keep building up with ever higher demand for pension protection with an increasingly older society in which more and more people will live in urban areas away from their original place of residence. The old model of "reactive gradualism" as employed in past fiscal reforms should be replaced with a more systematic reform framework (Wong and Bird 2005). But the central government has yet to succeed in adopting a comprehensive and concerted approach to tackle the challenges bound to emerge with these trends when reforming the pension system and the entire fiscal system at large (Wong 2013a, 2017).

Lots of promises were made for the workers and staff with rather generous benefits when the pension system was first set up in early 1950s, as can be seen in the comprehensive provisions spelled out in the "Labour Insurance Regulations of the PRC". But that was almost 70 years ago when the population was young and the life expectancy was short. The system was very affordable and the burden never materialised back then (World Bank 1997). With the much-increased demand for pension protection and the potential squeeze on available resources caused by continuing population ageing and urbanisation, both at unparalleled speed and scale, how to ensure the long-term sustainability of the pension system and spend the funds in the most efficient and effective way becomes very important issues.

Yet, after so many rounds of reforms, the administration of the pension system still remains highly fragmented and decentralised, with more than 2,000 separate local pools of pension funds and substantial regional variations in the specific rules of the system often determined at the discretion of local governments who are entrusted with the provision and management of pension services (Wong 2013a; Hussain 2007). The current state of the pension system is neither conducive to achieving efficient financing through a proper level of risk pooling and coordination, nor amenable to effective performance monitoring and management. Moreover, the dysfunctional incentive structures embedded in the fragmented fiscal system in general and pension system in particular hamper the efforts to achieve other national policy objectives that are important in the big picture of the general economic reform (Wong 2010). For example, as the unaffordability of pension contributions for most migrant workers has been exploited by the local governments to create barriers to limit migrant workers' access to urban public services, the fragmented pension administration hinders inclusive urbanisation, which is one of the goals set by the central government in its the New-Type Urbanisation Plan (CPC Central Committee and State Council 2014).

Reassuringly, the need for further pension reform has been acknowledged by the central government and the task has remained high on its agenda (CPC Central Committee 2015b; Central Government of the PRC 2016; National People's Congress 2021; Xi 2021). Some encouraging recent developments include earmarking 10% of the capital of the state-owned enterprises for the purpose of enriching the source of funding for the pension system, the establishment of a central adjustment fund to help fill the funding gaps for the provinces with deficits in their pension funds, and the plan to transfer the management of the collection of pension contributions to tax offices to strengthen and standardise collection management across the country (State Council 2017a, 2018; CPC Central Committee and State Council 2018). These latest reform measures are in the right direction and can be read as a display of capacity to boost public confidence in the government's ability to sustain the pension system financially. But they still remain incremental and reactive in nature. More substantial measures guided by a greater vision are required to consolidate and unify the highly fragmented pension system.

As revealed in the results of this research, a demographic tsunami may be on the horizon, but the fragmented pension system is brewing its own perfect storm from within. Compared to the population ageing, the issues stemming from the fragmentation pose a more insidious threat to pension sustainability in China. The retirement age reform alone can only provide a necessary but not sufficient condition for ensuring the system's long-run financial sustainability, even without considering the significant negative impact of the fragmentation. Problems of moral hazard such as noncompliance by local governments and challenges of adverse selection resulting from the administrative loopholes in the highly decentralised system, if left unchecked, are classic reasons why insurance policies including pension schemes go bankrupt. Therefore, if China wants to ensure the long-term sustainability of the pension system, it is imperative to take the pension reform to the next level by defragmenting the system. The possibility of the fertility cliff and the danger of the de facto bankruptcy brought by the population ageing further highlight the urgency to address the underlying cause of the many defects of the system that are damaging pension sustainability.

While designing the details of a nationally unified pension system for China is beyond the scope of this book, the new system should include several features as implied by the results of this study. First, it should embody a clear division of intergovernmental fiscal responsibility in pension financing as a starting point to address the issue of moral hazard. Second, the financial status of the pension system should be reported based on accrual accounting with annually updated projections of future cashflows and actuarial valuation of liabilities. Third, it should treat migrant workers fairly by removing the regressive contribution rules and making the accrued benefits fully portable to encourage their participation and promote inclusive and sustainable urbanisation as well as free flow of labour across provinces. Fourth, the new system should be free of the administrative loopholes that allow adverse selection to occur, such as those allowing local governments to decide who can join the system through the buy-in option and job agencies to create fictitious employment history for anyone able to afford the fees to obtain the entitlement to pension benefits from any cities at their choosing. Last but not least, the financing method of the new system should still be based on a PAYG system rather than a fully funded system in order to achieve an optimal welfare outcome across the generations as revealed by the quantitative analysis. This highlights the importance of ensuring a steady pace of development and productivity improvement in the economy, since the return of a PAYG system relies on the aggregate wage growth (Barr and Diamond 2006), which, in the context of China as a developing country, depends on unleashing the development potential from ensuring an inclusive and sustainable urbanisation process with free movement of labour.

The currently highly fragmented pension system is incompatible with these ingredients conducive for pension sustainability. The central government was aware of the suboptimal nature of the fragmented pension system and recognised the need to achieve higher levels of risk pooling as early as 1991 (State Council 1991). And the goal to eventually realise national-level pooling was reiterated in the Social Insurance Law promulgated in 2010 (National People's Congress 2010a). Yet, the fragmentation has persisted until now. The financing of the pension system was helped by the fiscal recovery that started in the late 1990s and a phase of rapid increase in total government revenues in the first decade of the 21st century, a fortuitous turn of event in the recent history of public finance in China where the central government was able to increase its spending in areas such as subsidising the pension system. Now with the slowdown of the economy in China, the government revenues can no longer sustain the double-digit growth. Combined with the mounting pressure from the ongoing rapid population ageing, the medium-to-long-term outlook of the sustainability of the pension system hangs in the balance. Thus, the circumstances under which the pension reform in China will proceed will become increasingly unforgiving. The writing was already on the wall, when the annual revenues of the UES excluding government subsidies started to be insufficient to cover the annual expenditures in 2015, as mentioned in Chapter 2. If the central government still cannot summon the resolve to revamp the system by fundamentally addressing the real elephant in the room that has been ignored for so long, it runs the risk of missing the critical window of opportunity that may not be there anymore. Fortunately, the technologies required to run a nationally unified pension system are long ready and getting even better year after year. The time is right to finally defragment the system.

References

51Shebao. 2018. 'China Enterprise Social Insurance White Paper 2018'. In Chinese. 51Shebao.

Anderson, Thomas, and Hans-Peter Kohler. 2013. 'Education Fever and the East Asian Fertility Puzzle: A Case Study of Low Fertility in South Korea'. Asian Population Studies 9 (2): 196–215. doi: 10.1080/17441730.2013.797293.

Barkan, Lenore. 1990. 'Chinese Old-Age Pension Reform: The Process Continues'. International Social Security Review 43 (4): 387–98. doi: 10.1111/j.1468-246X.1990. tb00871.x.

Barr, Nicholas, and Peter Diamond. 2006. 'The Economics of Pensions'. Oxford Review of Economic Policy 22 (1): 15–39. doi: 10.1093/oxrep/grj002.

BBC. 2015. 'Old-Age Insurance: China to Announce the Plan for Delaying Retirement'. BBC Chinese. 2015. http://www.bbc.com/zhongwen/simp/china/2015/03/150311_ ana_china_retirement_age.

Beijing Municipal Social Insurance Funds Management Centre. 2019. 'Notice on Unifying the Salary Bases and Contributions of Various Social Insurances in 2019'. In Chinese. http://www.beijing.gov.cn/zhengce/zhengcefagui/201907/t20190704_101119.html.

Bloomberg. 2013. 'China Eclipses U.S. as Biggest Trading Nation'. Bloomberg. 2013. http:// www.bloomberg.com/news/articles/2013-02-09/china-passes-u-s-to-become-the-world-s-biggest-trading-nation.

Breyer, Friedrich. 1989. 'On the Intergenerational Pareto Efficiency of Pay-as-You-Go Financed Pension Systems'. Journal of Institutional and Theoretical Economics (JITE) / Zeitschrift Für Die Gesamte Staatswissenschaft 145 (4): 643–58.

Bureau of Human Resources and Social Security of Beijing. 2017. 'Notice on Unifying the Salary Bases and Contributions of Various Social Insurances in Beijing in 2017'. In Chinese. http://www.gov.cn/xinwen/2017-11/23/content_5241805.htm.

———. 2018. 'Notice on Unifying the Salary Bases and Contributions of Various Social Insurances in 2018'. In Chinese. http://rsj.beijing.gov.cn/xxgk/tzgg/201912/ t20191207_953372.html.

Bureau of Human Resources and Social Security of Shanghai. 2017. 'The Maximum and Minimum Levels for the Permissible Salaries for Computing Social Security Contributions Determined for 2017 for Shanghai'. In Chinese. Website of Bureau of Human Resources and Social Security of Shanghai. 1 April 2017. http://www.12333sh.gov.cn/ newapp/07/201704/t20170401_1253445.shtml.

———. 2019. 'Basic Situation of Social Insurances in Shanghai in 2018'. In Chinese. http:// rsj.sh.gov.cn/201712333/xxgk/zdly/01/201906/t20190603_1297140.shtml.

Bureau of Human Resources and Social Security of Xiamen. 2018. 'The Lower Limit and Upper Limit for the Permissible Salary Base in Xiamen in 2018 Are 1,700 Yuan and 18,864 Yuan per Month Respectively'. In Chinese. https://www.12333ask.com/article/435.

Bureau of Social Insurances of Tongren. 2019. 'Announcement on Stopping the One-Off Payment of Basic Pension Insurance Fees for Urban Enterprise Employees'. In Chinese. http://rsj.trs.gov.cn/zwgk/xxgkml/gggs/201809/t20180921_3573539.html.

Bureau of Statistics of Beijing. 2017. 'How to Correctly Interpret the Meaning of the Average Wage for Urban Employees?' In Chinese. 29 August 2017. http://www.bjstats.gov.cn/tjzd/zswd/201708/t20170829_381272.html.

Bureau of Statistics of Guangdong. 2017. *Guangdong Statistical Yearbook 2017*. In Chinese. Beijing: China Statistics Press.

Bureau of Statistics of Heilongjiang. 2017. *Heilongjiang Statistical Yearbook 2017*. In Chinese. Beijing: China Statistics Press.

Bureau of Statistics of Hunan. 2017. *Hunan Statistical Yearbook 2017*. In Chinese. Beijing: China Statistics Press.

Bureau of Statistics of Shandong. 2017. *Shandong Statistical Yearbook 2017*. In Chinese. Beijing: China Statistics Press.

Bureau of Statistics of Shanghai. 2017. 'Living Conditions of Migrant Workers in Shanghai'. In Chinese. http://www.shanghai.gov.cn/nw2/nw2314/nw24651/nw42131/nw42178/u21aw1232783.html.

Bureau of Statistics of Sichuan. 2017. *Sichuan Statistical Yearbook 2017*. In Chinese. Beijing: China Statistics Press.

Cai, Fang. 2010. 'Demographic Transition, Demographic Dividend, and Lewis Turning Point in China'. *China Economic Journal* 3 (2): 107–19. doi: 10.1080/17538963.2010.511899.

Cai, Fang, Zhenwei Guo, and Meiyan Wang. 2016. 'New Urbanisation as a Driver of China's Growth'. In *China's New Sources of Economic Growth: Vol. 1*, edited by Ligang Song, Ross Garnaut, Fang Cai, and Lauren Johnston, 1st ed. ANU Press. https://doi.org/10.22459/CNSEG.07.2016.03.

Cai, Fang, and Wen Zhao. 2012. 'When Demographic Dividend Disappears: Growth Sustainability of China'. In *The Chinese Economy: A New Transition*, edited by Masahiko Aoki and Jinglian Wu, 75–90. London: Palgrave Macmillan UK. doi: 10.1057/9781137034298_5.

Cai, Yong, and Yuan Cheng. 2014. 'Pension Reform in China: Challenges and Opportunities'. *Journal of Economic Surveys* 28 (4): 636–51. doi: 10.1111/joes.12082.

Caijing. 2017. 'Three Trillion Yuan of Pension Subsidies in 15 Years: Actuarial Balancing of the Funds Required'. In Chinese. 2017. http://economy.caijing.com.cn/20170607/4281458.shtml.

CCTV. 2021. 'What Will Happen to the Manufacturing Industry in China When All the Young People Work in Food Delivery?' *CCTV*. 10 April 2021. https://news.cctv.com/2021/04/19/ARTI3Rtw9dZNiXGo9H1Dh4yk210419.shtml.

CCTV America. 2015. 'Chinese Fertility Rate Drops into "Low Fertility Trap"'. *CCTV America*. 2015. http://www.cctv-america.com/2015/01/05/chinese-fertility-rate-drops-into-low-fertility-trap.

Central Government of the PRC. 2016. 'Outline of the 13th Five-Year Plan for Economic and Social Development of the People's Republic of China'. In Chinese. http://www.gov.cn/xinwen/2016-03/17/content_5054992.htm

Chen, Qian. 2013. 'Analysis on Problems and Countermeasures of Basic Pension Insurance of Urban Workers Payment System in China'. In Chinese. *Scientific Research on Aging* 1 (4): 48–56.

Chen, Rong, Ping Xu, Peipei Song, Meifeng Wang, and Jiangjiang He. 2019. 'China Has Faster Pace than Japan in Population Aging in Next 25 Years'. *Bioscience Trends* 13 (4): 287–91. doi: 10.5582/bst.2019.01213.

Chen, Tianhong, and John A. Turner. 2015. 'Fragmentation in Social Security Old-Age Benefit Provision in China'. *Journal of Aging & Social Policy* 27 (2): 107–22. doi: 10.1080/08959420.2014.977647.

Chen, Vivian. 2004. *A Macro Analysis of China Pension Pooling System: Incentive Issues and Financial Problem. International Conference on Pensions in Asia: Incentives, Compliance and Their Role in Retirement.* Tokyo, Japan: Hitotsubashi University.

Chengdu Municipal People's Government. 2019. *Applying for Residence Permit.* In Chinese. Website of Chengdu Municipal People's Government. 26 March 2019. http://www.chengdu.gov.cn/chengdu/smfw/bljzz.shtml.

China Daily. 2015. 'Nation Must Be Alert to Middle-Income Trap'. *China Daily USA.* 2015. http://usa.chinadaily.com.cn/epaper/2015-04/28/content_20568239.htm.

Chongqing Daily. 2017. 'The Average Monthly Wage of Employees in Chongqing Reached 5616 Yuan Last Year'. In Chinese. *Chongqing Daily,* 17 June 2017. http://cq.people.com.cn/n2/2017/0617/c367668-30340343.html.

Chongqing Social Security Online. 2017. 'Minimum Wage for Chongqing in 2017'. In Chinese. *Chongqing Social Security Online.* 13 September 2017. http://chongqing.chashebao.com/ziliao/18016.html.

CPC Central Committee. 1993. 'Decision of the Central Committee of the Communist Party of China on Some Issues Concerning the Establishment of the Socialist Market Economy'. In Chinese. http://www.people.com.cn/item/20years/newfiles/b1080.html

———. 2003. 'Decision of the Central Committee of the Communist Party of China on Some Issues Concerning the Improvement of the Socialist Market Economy'. In Chinese.

———. 2004. 'Decision of the Central Committee of the Communist Party of China on Strengthening the Party's Governance Capability'. In Chinese.

———. 2005. 'Proposal of the Central Committee of the Communist Party of China on the Formulation of the 11th Five-Year Plan'. In Chinese.

———. 2013. 'Decision of the Central Committee of the Communist Party of China on Some Major Issues Concerning Comprehensively Deepening the Reform'. In Chinese. http://www.gov.cn/jrzg/2013-11/15/content_2528179.htm.

———. 2015a. 'Communique of the Fifth Plenary Session of the 18th CPC Central Committee'. In Chinese. http://www.caixin.com/2015-10-29/100867990_all.html#page2.

———. 2015b. 'Opinion of the Central Committee of the Communist Party of China on Setting the 13th Five-Year Plan for Economic and Social Development of the People's Republic of China'. In Chinese.

CPC Central Committee and State Council. 2014. 'National New-Type Urbanisation Plan (2014–2020)'. In Chinese. http://www.gov.cn/zhengce/2014-03/16/content_2640075.htm.

———. 2018. 'Reform Plan for the Collection and Management System for State Tax and Local Tax'. In Chinese. http://www.xinhuanet.com/politics/2018-07/20/c_1123156533.htm.

Deng, Quheng, and Shi Li. 2010. 'Wage Structures and Inequality Among Local and Migrant Workers in Urban China'. In *The Great Migration,* edited by Xin Meng, Chris Manning, Li Shi, and Tadjuddin Effendi, 74–92. Cheltenham: Edward Elgar Publishing. doi: 10.4337/9781781000724.00012.

Department of Human Resources and Social Security of Guangdong and Department of Finance of Guangdong. 2018. 'Notice on Further Regulating the Policy on Payment for Participating in the Basic Pension Insurance for Enterprise Employees'. In Chinese. http://szsi.sz.gov.cn/sbjxxgk/zcfggfxwj/zcfg/sbzs/201810/t20181019_14302673.htm.

Department of Human Resources and Social Security of Hunan and Department of Finance of Hunan. 2017. 'Notice on Properly Handling Issues Concerning the Payment of Basic Pension Insurance Fees for Enterprise Employees'. In Chinese. http://www.yiyang.gov.cn/yiyang/2/3/4/content_366045.html.

d'Haene, Yannick, and Pascale Emile. 1994. 'China: Reforming the Social Security System'. *International Social Security Review* 47 (2): 77–85. doi: 10.1111/j.1468-246X.1994.tb00402.x.

Diamond, Peter A. 1977. 'A Framework for Social Security Analysis'. *Journal of Public Economics* 8 (3): 275–98. doi: 10.1016/0047-2727(77)90002-0.

Dixon, John. 1981. *The Chinese Welfare System, 1949–1979.* New York: Praeger Publishers.

Dong, Keyong, and Xiangfeng Ye. 2003. 'Social Security System Reform in China'. *China Economic Review* 14 (4): 417–25. doi: 10.1016/j.chieco.2003.09.012.

Dorfman, Mark C., Robert Holzmann, Philip O'Keefe, Dewen Wang, Yvonne Sin, and Richard Hinz. 2013. *China's Pension System: A Vision.* The World Bank. https://doi.org/10.1596/978-0-8213-9540-0.

Economic Daily. 2018. 'The Urbanisation Rate by Household Registration Reached 42.35% in China in 2017: To Accelerate Turning Migrant Workers into Urban Residents'. In Chinese. *Economic Daily*, 12 April 2018. http://www.ce.cn/xwzx/gnsz/gdxw/201804/12/t20180412_28805119.shtml.

Elliott, Larry. 2015. 'China May Find Raising the Birth Rate Is No Simple Matter'. *The Guardian.* 2015. https://www.theguardian.com/world/2015/oct/29/china-may-find-raising-the-birth-rate-is-no-simple-matter.

Fanti, Luciano, and Luca Gori. 2012. 'Fertility and PAYG Pensions in the Overlapping Generations Model'. *Journal of Population Economics* 25 (3): 955–61. doi: 10.1007/s00148-011-0359-7.

Feldstein, Martin. 1998. 'Social Security Pension Reform in China'. Working Paper 6794. National Bureau of Economic Research. doi:10.3386/w6794.

Feng, Jin. 2016. *Sustainable Pension Level: From the Perspectives of Globalisation, Urbanisation and Population Ageing.* In Chinese. Beijing: China CITIC Press.

Feng, Xu. 2018a. 'Employees Want Cash and Employers Want to Save Money: How to Solve the Problem of "Refusal to Participate in Social Insurances"'. In Chinese. *Dongnan Net.* 2018. http://fjnews.fjsen.com/2018-04/28/content_20988635.htm.

——— 2018b. 'Working for 12 Years in Jinjiang without Being Enrolled in Social Insurances: Employee Cannot Claim Huge Medical Bills from Medical Insurance'. In Chinese. *Dongnan Net.* 2018. http://fjnews.fjsen.com/2018-04/28/content_20987328.htm.

Fock, Achim, and Christine Wong. 2008. 'Financing Rural Development for a Harmonious Society in China: Recent Reforms in Public Finance and Their Prospects'. Policy Research Working Paper No. WPS 4693. Washington, DC: World Bank.

Galor, Oded. 2012. 'The Demographic Transition: Causes and Consequences'. *Cliometrica* 6 (1): 1–28. doi: 10.1007/s11698-011-0062-7.

Ghilarducci, Teresa, and Kevin Terry. 1999. 'Scale Economies in Union Pension Plan Administration: 1981–1993'. *Industrial Relations: A Journal of Economy and Society* 38 (1): 11–17. doi: 10.1111/0019-8676.00107.

Hainan Provincial Committee of Jiu San Society. 2016. '*Suggestions on Improving the Issue of Social Insurance Discontinuation by Migrant Workers*'. In Chinese. People's Government of Hainan Province. 2016. http://www.hainan.gov.cn/zxtadata-7086.html.

Hangzhou Municipal People's Government. 2017. 'Notice of the General Office of the People's Government of Hangzhou on the Implementation of the New Residence Permit System'. In Chinese. https://zjjzzgl.zjsgat.gov.cn/zahlw/xxgkXx?nbbh=402882fa5b5bd990015b5be9ff150000.

Hosseini, Roozbeh. 2015. 'Adverse Selection in the Annuity Market and the Role for Social Security'. *Journal of Political Economy* 123 (4): 941–84. doi:10.1086/681593.

HSBC. 2017. *Global Report: The Value of Education-Higher and Higher.* London: HSBC Holdings plc.

Hu, Aidi. 1997. 'Reforming China's Social Security System: Facts and Perspectives'. *International Social Security Review* 50 (3): 45–65. doi: 10.1111/j.1468-246X.1997.tb01076.x.

Huang, Jiawen. 2019. 'Income Inequality, Distributive Justice Beliefs, and Happiness in China: Evidence from a Nationwide Survey'. *Social Indicators Research* 142 (1): 83–105. doi: 10.1007/s11205-018-1905-4.

Humblet, Martine, and Rosinda Silva. 2002. *Social Security: Standards for the XXIst Century.* Geneva: International Labour Office.

Hussain, Athar. 1994. 'Social Security in Present-Day China and Its Reform'. *American Economic Review* 84 (2): 276–80.

———— 2007. 'Social Security in Transition'. In *Paying for Progress in China: Public Finance, Human Welfare and Changing Patterns of Inequality*, edited by Vivienne Shue and Christine Wong, 96–116. Routledge Contemporary China Series 21. London: Routledge.

Insurance Association of China. 2015. *Insurance Association of China Releases the Report on Employees Pension Reserve Index in Large and Medium-Sized Cities in China in 2015.* In Chinese. http://www.iachina.cn/content_b834881c-9727-11e5-af37-2e68814cf9fe.html.

Ioannides, Stavros, and Klaus Nielsen, eds. 2007. *Economics and the Social Sciences: Boundaries, Interaction and Integration.* Cheltenham, Glos, UK; Northampton, MA: Edward Elgar.

James, Estelle. 2002. 'How Can China Solve Its Old-Age Security Problem? The Interaction between Pension, State Enterprise and Financial Market Reform'. *Journal of Pension Economics and Finance* 1 (1): 53–75. doi: 10.1017/S1474747202001026.

Jenkins, Rhys, Enrique Dussel Peters, and Mauricio Mesquita Moreira. 2008. 'The Impact of China on Latin America and the Caribbean'. *World Development* 36 (2): 235–53. doi: 10.1016/j.worlddev.2007.06.012.

Jin, Gang. 2010. 'The Analysis on the Status and Problems of China's Retirement Age and the Necessity of Extending Retirement Age'. In Chinese. *Social Security Studies* 2: 32–38.

Jin, Yongai. 2014. 'Low Fertility Trap: Theories, Facts and Implications'. In Chinese. *Population Research* 38 (1): 3–17.

Johnston, Lauren A. 2019. 'The Economic Demography Transition: Is China's "Not Rich, First Old" Circumstance a Barrier to Growth?' *Australian Economic Review* 52 (4): 406–26. doi: 10.1111/1467-8462.12325.

Lee, Leng. 2012. 'Decomposing Wage Differentials between Migrant Workers and Urban Workers in Urban China's Labor Markets'. *China Economic Review* 23 (2): 461–70. doi: 10.1016/j.chieco.2012.03.004.

Legal Daily. 2006. 'The Criminal Case Involving Social Insurance Funds in Shanghai'. In Chinese. *Legal Daily.* 2006. http://news.sina.com.cn/c/2006-12-29/183711917747.shtml.

Leung, Joe C.B. 2003. 'Social Security Reforms in China: Issues and Prospects'. *International Journal of Social Welfare* 12 (2): 73–85. doi: 10.1111/1468-2397.t01-1-00246.

Li, Keqiang. 2015. 'Report on the Work of the Government (2015)'. *Chinese Central Government.* 2015. http://english.gov.cn/archive/publications/2015/03/05/content_281475066179954.htm.

Li, Miao. 2016. 'Don't Let the Migrant Children Schools Sink in "Low Quality"'. In Chinese. *China Youth Daily*, 22 February 2016. http://zqb1.cyol.com/html/2016-02/22/nw.D110000zgqnb_20160222_1-10.htm.

Li, Tieying. 2000. 'Establish a Social Security System with Chinese Features'. In *Social Security Reform: Options for China*, edited by Jason Z. Yin, Shuanglin Lin, and David F. Gates, 39–50. Singapore: World Scientific.

Lieberthal, Kenneth. 2004. *Governing China: From Revolution through Reform.* 2nd ed. New York: W. W. Norton.

Liu, Wan. 2013. 'Does Late Retirement Surely Damage the Retired Interest? An Investigation on the Pension Wealth of Various Retirement Ages of Urban Employees'. In Chinese. *Economic Review* 4: 27–36.

Lou, Jiwei. 2015. 'Lou Jiwei's Speech at Tsinghua University: China May Fall into the Middle Income Trap'. In Chinese. 2015. http://finance.sina.com.cn/china/20150501/135822089571.shtml.

———— 2019. 'Understanding the Budgetary Issues Based on the National Conditions'. In Chinese. *Public Finance Research* 5: 3–6.

Lu, B., and J. Piggott. 2015. 'Meeting the Migrant Pension Challenge in China'. *CESifo Economic Studies* 61 (2): 438–64. doi: 10.1093/cesifo/ifu017.

Lutz, Wolfgang, Vegard Skirbekk, and Maria Rita Testa. 2006. 'The Low-Fertility Trap Hypothesis: Forces That May Lead to Further Postponement and Fewer Births in Europe'. *Vienna Yearbook of Population Research* 4: 167–92.

Mai, Yinhua, Xiujian Peng, and Wei Chen. 2013. 'How Fast Is the Population Ageing in China?' *Asian Population Studies* 9 (2): 216–39. doi: 10.1080/17441730.2013.797295.

McCurry, Justin, and Julia Kollewe. 2011. 'China Overtakes Japan as World's Second-Largest Economy'. *The Guardian*. 2011. http://www.theguardian.com/business/2011/feb/14/china-second-largest-economy.

Meijdam, Lex, and Harrie Verbon. 1997. 'Aging and Public Pensions in an Overlapping Generations Model'. *Oxford Economic Papers* 49 (1): 29–42. doi: 10.1093/oxfordjournals.oep.a028595.

Ministry of Finance. 1969. 'Opinions on the Reform of Several Systems in the Financial Work of the State-Owned Enterprises'. In Chinese.

———. 2015. *Finance Yearbook of China 2015*. In Chinese. Beijing: Ministry of Finance.

———. 2019a. '2018 Final Account of National General Public Budget Revenue'. In Chinese. http://yss.mof.gov.cn/2018czjs/201907/t20190718_3303107.htm.

———. 2019b. '2018 Final Account of National Government-Managed Funds Revenue'. In Chinese. http://yss.mof.gov.cn/2018czjs/201907/t20190718_3303321.htm.

———. 2019c. '2018 Final Account of National Social Insurance Funds Revenue'. In Chinese. http://yss.mof.gov.cn/2018czjs/201907/t20190718_3303339.htm.

Ministry of Human Resources and Social Security. 2011a. '2010 Statistical Bulletin on the Development of Human Resources and Social Security'. In Chinese.

———. 2011b. *The Old-Age Social Insurance*. In Chinese. Beijing: China Labour and Social Security Publishing House.

———. 2012. 'Interpretation of the Social Insurance Law of the People's Republic of China (11)'. In Chinese. http://www.mohrss.gov.cn/SYrlzyhshbzb/rdzt/syshehuibaoxianfa/bxffaguijijiedu/201208/t20120806_28572.htm.

———. 2016a. '2015 Statistical Bulletin on the Development of Human Resources and Social Security'. In Chinese. http://www.mohrss.gov.cn/SYrlzyhshbzb/zwgk/szrs/tjgb/201805/W020180521566641377904.pdf.

———. 2016b. *Annual Report on China's Social Insurance Development in 2015*. Beijing: China Human Resources and Social Security Publishing Group.

———. 2017. 'MOHRSS: Pension Benefits Continue to Increase at an Average Rate of 5.5% in 2017'. In Chinese. http://www.mohrss.gov.cn/zcyjs/gongzuodongtai/201707/t20170729_274838.html.

———. 2018. '2017 Statistical Bulletin on the Development of Human Resources and Social Security'. In Chinese. http://www.mohrss.gov.cn/SYrlzyhshbzb/zwgk/szrs/tjgb/201805/t20180521_294287.html.

———. 2019. '2018 Statistical Bulletin on the Development of Human Resources and Social Security'. In Chinese. http://www.mohrss.gov.cn/SYrlzyhshbzb/zwgk/szrs/tjgb/201906/t20190611_320429.html.

———. 2020. '2019 Statistical Bulletin on the Development of Human Resources and Social Security'. http://www.mohrss.gov.cn/SYrlzyhshbzb/zwgk/szrs/tjgb/202006/t20200608_375774.html.

Ministry of Human Resources and Social Security and Ministry of Finance. 2009. 'Interim Measures for the Transfer of Basic Old-Age Insurance for Employees in Urban Enterprises'. In Chinese. http://www.gov.cn/zwgk/2009-12/29/content_1499072.htm.

———. 2010. 'Opinions on Resolving the Remaining Issues Such as Basic Old-Age Social Security for Retirees of Uncovered Collective Enterprises'. In Chinese. http://www.mohrss.gov.cn/gkml/zcfg/gfxwj/201702/t20170217_266343.html.

———. 2016a. 'Notice on Further Strengthening the Management of Revenues and Expenditures of Basic Pension Insurance Funds for Enterprise Employees'. In Chinese. https://rst.hebei.gov.cn/a/zhengce/zhengce/shengting/2019/0104/7059.html.

———. 2016b. 'Notice on the Temporary Reduction of Social Insurance Contribution Rates'. In Chinese. http://www.mof.gov.cn/zhengwuxinxi/zhengcefabu/201604/t20160421_1959347.htm.

———. 2018. 'Notice on the Extension for the Temporary Reduction of Social Insurance Contribution Rates'. In Chinese. http://www.mof.gov.cn/zhengwuxinxi/caizhengxinwen/201804/t20180428_2880642.htm.

———. 2021. 'Notice on Adjustments of the Basic Pesion Benefits for Retirees in 2021'. In Chinese. http://www.gov.cn/zhengce/zhengceku/2021-04/17/content_5600226.htm.

Ministry of Human Resources and Social Security, Ministry of Finance, and State Taxation Administration. 2020. 'Notice on the Temporary Reduction and Exemption of Enterprise Social Insurance Fees'. http://www.mohrss.gov.cn/gkml/zcfg/gfxwj/202002/t20200221_360350.html.

Ministry of Labour. 1953. 'Draft Amendments to the Rules for Implementation of Labour Insurance Regulations of the People's Republic of China'. In Chinese.

———. 1993. 'Notification by the Ministry of Labour on the Work for the Pilot Reform of the Rules of Benefits Determination for the Basic Pension'. In Chinese.

Ministry of Labour and Social Security. 2001. 'Notification on Some Issues Concerning Improving the Policy of Basic Old-Age Insurance for Urban Workers'. In Chinese.

———. 2006. '2005 Statistical Bulletin on the Development of Labour and Social Security'. In Chinese.

Montes, Manuel, and Manuel Riesco. 2018. 'Pensions: Risking the Twentieth Century's Biggest Social Gain'. DOC Research Institute. 16 February 2018. https://doc-research.org/2018/02/chile-pensions-2/.

Mulligan, Casey, and Xavier Sala-i-Martin. 1999. *Social Security in Theory and Practice (II): Efficiency Theories, Narrative Theories, and Implications for Reform*. w7119. Cambridge, MA: National Bureau of Economic Research. doi: 10.3386/w7119.

National Audit Office. 2012. 'Audit Results for Social Security Funds across the Country'. In Chinese. http://www.gov.cn/zwgk/2012-08/02/content_2196871.htm.

National Bureau of Statistics. 1997. *China Labour Statistical Yearbook 1997*. In Chinese. Beijing: China Statistics Press.

———. 2011. 'Main Data Bulletin of the Sixth National Population Census in 2010'. In Chinese. National Bureau of Statistics of China.

———. 2012. *China Statistical Yearbook 2012*. In Chinese. Beijing: China Statistics Press.

———. 2013. 'Survey Report on the Monitoring of Migrant Workers in 2012'. In Chinese. http://www.stats.gov.cn/tjsj/zxfb/201305/t20130527_12978.html.

———. 2015. 'Survey Report on the Monitoring of Migrant Workers in 2014'. In Chinese. http://www.stats.gov.cn/tjsj/zxfb/201504/t20150429_797821.html.

———. 2016a. *China Labour Statistical Yearbook 2016*. In Chinese. Beijing: China Statistics Press.

———. 2016b. *China Statistical Yearbook 2016*. In Chinese. Beijing: China Statistics Press.

———. 2017. *China Statistical Yearbook 2017*. In Chinese. Beijing: China Statistics Press.

———. 2018a. *China Labour Statistical Yearbook 2018*. In Chinese. Beijing: China Statistics Press.

———. 2018b. *China Statistical Yearbook 2018*. In Chinese. Beijing: China Statistics Press.

———. 2018c. 'Survey Report on the Monitoring of Migrant Workers in 2017'. In Chinese. http://www.stats.gov.cn/tjsj/zxfb/201804/t20180427_1596389.html.

———. 2019a. *China Statistical Yearbook 2019*. In Chinese. Beijing: China Statistics Press.

———. 2019b. 'Survey Report on the Monitoring of Migrant Workers in 2018'. In Chinese. http://www.stats.gov.cn/tjsj/zxfb/201904/t20190429_1662268.html.

National Bureau of Statistics of China. 2018. 'Incomes and Expenditures for Residents in 2017'. In Chinese. http://www.stats.gov.cn/tjsj/zxfb/201801/t20180118_1574931.html.

National Business Daily. 2021. 'v17 Consecutive Increases! Pension benefits to increase by 4.5% in 2021 with two types of retirees likely to enjoy higher increases'. *National Business Daily*. 2021. http://www.nbd.com.cn/articles/2021-04-16/1702243.html.

National Development and Reform Commission. 2020a. 'Implementation Plan for Promoting the "Utilising Cloud Computing" Action and Nurturing the New Economy'. In Chinese. http://www.gov.cn/zhengce/zhengceku/2020-04/10/content_5501163.htm.

———. 2020b. 'Opinions on Supporting the Healthy Development of New Industries and New Business Models to Activate the Consumption Markets and Increase Employment'. In Chinese. https://www.ndrc.gov.cn/xxgk/zcfb/tz/202007/t20200715_1233793.html.

National Health and Family Planning Commission. 2016. *2016 Report on China's Migrant Population Development*. In Chinese. Beijing: China Population Press.

National People's Congress. 1994. 'Labour Law of the People's Republic of China'. In Chinese. http://www.gov.cn/banshi/2005-05/25/content_905.htm.

———. 2010a. 'Social Insurance Law of the People's Republic of China'. In Chinese.

———. 2010b. '"Two Assurances" and "Three Guarantee Lines"'. In Chinese. Website of National People's Congress. 2010. http://www.npc.gov.cn/zgrdw/npc/xinwen/2010-02/25/content_1543960.htm.

———. 2015. 'Population and Family Planning Law of the People's Republic of China'. In Chinese. http://www.gov.cn/xinwen/2015-12/28/content_5028414.htm.

———. 2021. 'The Fourteenth Five-Year Plan for National Economic and Social Development and the Outline of Long-Term Goals for 2035 of the People's Republic of China'. In Chinese. http://www.xinhuanet.com/2021-03/13/c_1127205564.htm.

OECD. 2013. *Pensions at a Glance 2013: OECD and G20 Indicators*. OECD Pensions at a Glance. OECD. doi: 10.1787/pension_glance-2013-en.

———. 2017. *Pensions at a Glance 2017: OECD and G20 Indicators*. OECD Pensions at a Glance. OECD. doi: 10.1787/pension_glance-2017-en.

Oksanen, Heikki. 2012. 'China: Pension Reform for an Aging Economy'. In *Nonfinancial Defined Contribution Pension Schemes in a Changing Pension World*, edited by Robert Holzmann and Edward Palmer, 213–258. Washington, D.C.: The World Bank. doi: 10.1596/9780821388488_CH07.

Pakrashi, Debayan, and Paul Frijters. 2017. 'Migration and Discrimination in Urban China: A Decomposition Approach'. *Review of Income and Wealth* 63 (4): 821–40. doi:10.1111/roiw.12245.

Pan, Yiquan, and Youqian Chen. 2009. 'The Bottom of the Society and the Politics of Resistance: A Study on the Survival and Social Ecology of Migrant Workers'. In Chinese. *Forward Position* 2: 109–12.

Pecchenino, Rowena A., and Patricia S. Pollard. 1997. 'The Effects of Annuities, Bequests, and Aging in an Overlapping Generations Model of Endogenous Growth'. *The Economic Journal* 107 (440): 26–46.

People's Daily 2010. 'How Should China Cope with Population Ageing?' In Chinese. *People's Daily Online*. 2010. http://paper.people.com.cn/rmrbhwb/html/2010-10/11/content_640839.htm.

——— 2015. 'Employee of Social Security Bureau of Yangchun Received Bribery Amounting to Almost Three Million Yuan'. In Chinese. *People's Daily Online*. 2015. http://cpc.people.com.cn/n/2015/0127/c87228-26454449.html.

——— 2016. 'The Total Number of Civil Servants in China First Disclosed: 7.167 Million in Total by the End of 2015'. In Chinese. *People's Daily Online*. 2016. http://politics.people.com.cn/n1/2016/0621/c1001-28464163.html.

———— 2018. 'Report: Number of Couriers Personnel Exceeded 3 Million with Average Monthly Wage at 6,200 Yuan'. In Chinese. *People's Daily Online*. 9 August 2018. http://industry.people.com.cn/n1/2018/0809/c413883-30219367.html.

People's Government of Beijing. 2018. 'Survey Report on the Monitoring of Migrant Workers in Beijing in 2017'. In Chinese. http://www.beijing.gov.cn/gongkai/shuju/sjjd/t1570251.htm.

People's Government of Chongyi County. 2016. 'National Poverty Standard by Year'. In Chinese. 2016. http://xxgk.chongyi.gov.cn/bmgkxx/qcz/gzdt/gggs/201612/t20161221_295794.htm.

People's Government of Guangdong. 1993. 'Provisional Regulations on the Old-Age Social Insurance for Employees in Guangdong'. In Chinese. http://sft.gd.gov.cn/sfw/gov_gk/gdsrmzfgz/content/post_3162774.html.

People's Government of Hunan. 2019. 'Notice of the People's Government of Hunan on Improving the Provincial-Level Coordination System for Basic Old-Age Insurance for Enterprise Employees'. In Chinese. http://www.hunan.gov.cn/xxgk/wjk/szfwj/201906/t20190613_5356947.html.

People's Government of Jixi County. 2019. 'Enrolment Size for the Urban Employees Pension Insurance in August 2019'. In Chinese. http://www.cnjx.gov.cn/openness/detail/content/5d7ef5e04b8edee0819bf83c.html.

People's Government of Shanghai. 1994. 'Measures for the Old-Age Insurance for Urban Employees in Shanghai'. In Chinese. http://rsj.sh.gov.cn/txgszfgz_17262/20200617/t0035_1388504.html.

————. 2014. 'Employees' Social Insurance Contribution Standard for Shanghai in 2014'. In Chinese. http://www.shanghai.gov.cn/nw2/nw2314/nw2319/nw23195/nw23200/u26aw42670.html.

People's Government of Tibet Autonomous Region. 2006. 'Circular of the People's Government of Tibet Autonomous Region on the Issuance of the "Implementation Plan for Improving the Basic Old-Age Insurance for Urban Employees in Tibet Autonomous Region"'. In Chinese. https://m12333.cn/policy/pmca.html.

————. 2007. 'Circular of the People's Government of Tibet Autonomous Region on the Issuance of the "Detailed Implementation Rules for Improving the Basic Old-Age Insurance for Urban Employees in Tibet Autonomous Region"'. In Chinese. https://m12333.cn/policy/ppes.html.

Remington, Thomas F. 2018. 'Institutional Change in Authoritarian Regimes: Pension Reform in Russia and China'. *Problems of Post-Communism* 66 (5): 301–314. doi: 10.1080/10758216.2018.1450154.

Runde, Daniel F. 2020. 'Will Many Developing Countries Get Old Before They Get Rich?' Center for Strategic and International Studies. 31 March 2020. https://www.csis.org/analysis/will-many-developing-countries-get-old-they-get-rich.

Saich, Tony. 2004. *Governance and Politics of China*. 2nd ed. Basingstoke, Hampshire, UK; New York: Palgrave Macmillan.

Salditt, Felix, Peter Whiteford, and Willem Adema. 2007. 'Pension Reform in China: Progress and Prospects'. *OECD Social, Employment and Migration Working Papers*, no. 53.

Schiermeier, Quirin. 2015. 'China's Birth Rate Won't Be Dramatically Affected by End of One-Child Policy'. *Nature News*. 2015. http://www.nature.com/news/china-s-birth-rate-won-t-be-dramatically-affected-by-end-of-one-child-policy-1.18687.

Shue, Vivienne, and Christine Wong. 2007. 'Introduction: Is China Moving to a More Equitable Development Strategy?' In *Paying for Progress in China: Public Finance, Human Welfare and Changing Patterns of Inequality*, edited by Vivienne Shue and Christine Wong, 1–11. Routledge Contemporary China Series 21. London: Routledge.

Sin, Yvonne, and Xiaoqing Yu. 2005. 'China - Pension Liabilities and Reform Options for Old Age Insurance'. Working Paper Series No. 2005-1. Washington, DC: The World Bank.

Sina News. 2016. 'Plan to Be Launched to Convert 100 Million Migrants to Urban Residents'. In Chinese. *Sina News*. 2016. http://news.sina.com.cn/o/2016-04-17/doc-ifxrizpp1464296.shtml.

———. 2019. 'Leaders of State-Owned Enterprises in Chongqing Showing Their "Pay Slips": Take a Look at Their Annual Salary Level'. In Chinese. *Sina News*. 13 February 2019. https://news.sina.com.cn/c/2019-02-13/doc-ihrfqzka5295574.shtml.

Social Security Administration. 2017. 'Annual Statistical Supplement to the Social Security Bulletin, 2016'. 2017. https://www.ssa.gov/policy/docs/statcomps/supplement/2016/supplement16.pdf.

———. 2019. 'Social Security Administrative Expenses'. Social Security Administration. 2019. https://www.ssa.gov/oact/STATS/admin.html.

Song, Zheng, Kjetil Storesletten, Yikai Wang, and Fabrizio Zilibotti. 2015. 'Sharing High Growth across Generations: Pensions and Demographic Transition in China'. *American Economic Journal: Macroeconomics* 7 (2): 1–39. doi: 10.1257/mac.20130322.

Standing Committee of Shenzhen People's Congress. 2014. 'Regulations on Residence Permit of Shenzhen Special Economic Zone'. In Chinese.

Standing, Guy. 2011. *The Precariat: The New Dangerous Class*. London, UK; New York, NY: Bloomsbury.

State Council. 1951. 'Labour Insurance Regulations of the People's Republic of China'. In Chinese. http://www.gov.cn/zhengce/2020-12/25/content_5574196.htm.

———. 1955. 'Interim Measures for Retirement of Staff of State Organs'. In Chinese. http://fgcx.bjcourt.gov.cn:4601/law?fn=chl521s429.txt&truetag=809&titles=&contents=&dbt=chl.

———. 1958. 'Provisional Regulations on the Retirement of Workers and Staff by the State Council'. In Chinese. http://www.npc.gov.cn/wxzl/gongbao/2000-12/09/content_5009558.htm.

———. 1978. 'State Council Circular on Issuing "State Council Temporary Measures on Providing for Old, Weak, Sick and Handicapped Cadres" and "State Council Temporary Measures on Workers' Retirement and Resignation"'. In Chinese. http://www.npc.gov.cn/wxzl/wxzl/2000-12/07/content_9548.htm

———. 1986. 'Circular of the State Council on Issuing the Four Provisions for the Reform of the Labour System'. In Chinese. http://www.gov.cn/zhengce/content/2012-09/21/content_7444.htm.

———. 1991. 'Decision on the Reform of the Old-Age Insurance System for the Employees of Enterprises'. In Chinese. http://www.reformdata.org/1991/0626/6677.shtml.

———. 1995. 'Notification on Deepening the Reform of the Old-Age Insurance System for Employees in Enterprises'. In Chinese. http://www.reformdata.org/1995/0301/4370.shtml.

———. 1997. 'Decision of the State Council on the Establishment of a Unified Basic Old-Age Insurance System for Employees of Enterprises'. In Chinese. http://www.people.com.cn/GB/shizheng/252/7486/7498/20020228/675965.html.

———. 2000. 'Notification of the State Council on Issuing the Pilot Programme for the Improvement of the Urban Social Security System'. In Chinese. http://www.gov.cn/xxgk/pub/govpublic/mrlm/201011/t20101112_62507.html.

———. 2001. 'Notice of the General Office of the State Council on the Prohibition of Local Governments from Raising the Level of Benefits for the Basic Pension Insurance for Enterprises on Their Own'. In Chinese. http://www.gov.cn/gongbao/content/2001/content_60975.htm.

———. 2005. 'Decision of the State Council on Improving the Basic Old-Age Insurance System for Employees of Enterprises'. In Chinese. http://www.gov.cn/zwgk/2005-12/14/content_127311.htm.

———. 2007. 'Social Insurance Law of the People's Republic of China (Draft)'. In Chinese. http://www.haikou.gov.cn/hdjl/myzji/201103/t20110306_186997.html.

————. 2008. 'State Council Measures for the Pilot Reform on the Reform of the Old-Age Insurance System for Public Institutions Employees'. In Chinese. http://www.reformdata.org/2008/0314/15748.shtml.

————. 2009. 'Guidance of the State Council on the Pilot Reform of the New Rural Old-Age Social Insurance'. In Chinese. http://www.gov.cn/zhengce/content/2009-09/04/content_7280.htm.

————. 2011. 'Guidance of the State Council on the Pilot Reform of Old-Age Social Insurance for Urban Residents'. In Chinese. http://www.gov.cn/zwgk/2011-06/13/content_1882801.htm.

————. 2014a. 'Circular of the State Council on the Approval and Transmission of the Reform Plan by the Ministry of Finance for the Accrual Government Comprehensive Financial Reporting System'. In Chinese. http://www.gov.cn/zhengce/content/2014-12/31/content_9372.htm.

————. 2014b. 'Opinion of the State Council on the Establishment of a Unified Basic Old-Age Insurance System for Urban and Rural Residents'. In Chinese. http://www.gov.cn/zhengce/content/2014-02/26/content_8656.htm.

————. 2015a. 'Decision of the State Council on the Reform of the Old-Age Insurance System for Civil Servants and Public Institutions Employees'. In Chinese. http://www.gov.cn/zhengce/content/2015-01/14/content_9394.htm.

————. 2015b. 'Interim Regulations on Residence Permits'. In Chinese. http://www.gov.cn/zhengce/content/2015-12/12/content_10398.htm.

————. 2017a. 'Notification of the State Council on Issuing the Implementation Plan of Transferring Some State-Owned Capital to Enrich Social Security Funds'. In Chinese. http://www.gov.cn/zhengce/content/2017-11/18/content_5240652.htm.

————. 2017b. 'Circular of the State Council on the Issuance of the National Population Development Plan (2016–2030)'. In Chinese. http://www.gov.cn/zhengce/content/2017-01/25/content_5163309.htm.

————. 2018. 'Notice of the State Council on the Establishment of the Central Adjustment System of the Basic Pension Insurance Fund for Enterprise Employees'. In Chinese. http://www.gov.cn/zhengce/content/2018-06/13/content_5298277.htm.

————. 2019. 'Circular of the General Office of the State Council on Issuing a Comprehensive Plan for Reducing Social Insurance Contribution Rates'. In Chinese. http://www.gov.cn/zhengce/content/2019-04/04/content_5379629.htm.

State Council Information Office. 2004. 'The Situation and Policies of Social Security in China'. In Chinese. http://www.gov.cn/gongbao/content/2004/content_62994.htm.

————. 2021. '*The State Council Information Office Held Press Conference on the Situation about Employment and Social Security*'. In Chinese. 2021. http://www.gov.cn/xinwen/2021-03/01/content_5589524.htm.

State Information Center. 2021. 'The 2021 Report on the Sharing Economy in China'. State Information Center. https://www.ndrc.gov.cn/xxgk/jd/wsdwhfz/202102/t20210222_1267536.html.

Sun, Jianpeng, and Qun Jiang. 2016. Analysis of the Causes and Solution for the Problem of "Empty Individual Accounts" in the Chinse Pension System'. In Chinese. *Human Resources Development* 6: 161–162.

Swider, Sarah Christine. 2015. *Building China: Informal Work and the New Precariat*. Ithaca; London: ILR Press, an imprint of Cornell University Press.

Tan, Kong Yam, Tilak Abeysinghe, and Khee Giap Tan. 2015. 'Shifting Drivers of Growth: Policy Implications for ASEAN-5'. *Asian Economic Papers* 14 (1): 157–73. doi: 10.1162/ASEP_a_00331.

Tan, Zhonghe, and Yongzhen Tan. 2016. *Risk Management Network: Supervision on Social Insurance Funds*. In Chinese. Beijing: China Democracy and Legal System Publishing House.

Tanner, Murray Scot. 2005. *Chinese Government Responses to Rising Social Unrest*. RAND Corporation Testimony Series. Santa Monica, CA: RAND Corporation.

Tao, Ran, and Zhigang Xu. 2007. 'Urbanization, Rural Land System and Social Security for Migrants in China'. *Journal of Development Studies*. doi:10.1080/00220380701526659.

The Beijing News. 2019. 'Former Employee of Department of Human Resources and Social Security of Hebei Stole over 70 Million Yuan of Pension Benefits'. In Chinese. *The Beijing News*. 2019. http://www.bjnews.com.cn/news/2019/12/28/667493.html.

The Economist. 2019. 'South Korea's Fertility Rate Falls to a Record Low'. *The Economist*. 2019. https://www.economist.com/graphic-detail/2019/08/30/south-koreas-fertility-rate-falls-to-a-record-low.

UNICEF. 2017. 'UNICEF Annual Report 2017: China'. UNICEF. https://www.unicef.org/about/annualreport/files/China_2017_COAR.pdf.

United Nations. 2015. 'World Population Prospects'. World Population Prospects. 2015. https://esa.un.org/unpd/wpp/.

———. 2019a. 'Percentage of Total Population by Broad Age Group, Both Sexes (per 100 Total Population)'. 2019. https://population.un.org/wpp/DataQuery/.

———. 2019b. 'Total Dependency Ratio ((Age 0-14 + Age 65+) / Age 15-64)'. 2019. https://population.un.org/wpp/Download/Standard/Population/.

———. 2019c. *World Population Prospects: The 2019 Revision*. http://data.un.org/Data.aspx?d=PopDiv&f=variableID%3A54.

Verbič, Miroslav, Boris Majcen, and Renger Van Nieuwkoop. 2006. 'Sustainability of the Slovenian Pension System: An Analysis with an Overlapping-Generations General Equilibrium Model'. *Eastern European Economics* 44 (4): 60–81. https://doi.org/10.2753/EEE0012-8775440403.

Wang, Lijian, Daniel Béland, and Sifeng Zhang. 2014a. 'Pension Fairness in China'. *China Economic Review* 28: 25–36. doi: 10.1016/j.chieco.2013.11.003.

——— 2014b. 'Pension Financing in China: Is There a Looming Crisis?' *China Economic Review* 30: 143–54. doi: 10.1016/j.chieco.2014.05.014.

Wang, Xiaojun. 2002. 'Analysis on the Financial Sustainability of Pension System in China'. In Chinese. *Market and Demographic Analysis* 8 (2): 26–29.

Weil, Philippe. 2008. 'Overlapping Generations: The First Jubilee'. HAL. https://EconPapers.repec.org/RePEc:hal:journl:hal-01022015.

West, Loraine A. 1999. 'Pension Reform in China: Preparing for the Future'. *The Journal of Development Studies* 35 (3): 153–83.

Williamson, John B., Meghan Price, and Ce Shen. 2012. 'Pension Policy in China, Singapore, and South Korea: An Assessment of the Potential Value of the Notional Defined Contribution Model'. *Journal of Aging Studies* 26 (1): 79–89. doi: 10.1016/j.jaging.2011.08.002.

Wong, Christine. 1992. 'Fiscal Reform and Local Industrialization: The Problematic Sequencing of Reform in Post-Mao China'. *Modern China* 18 (2): 197–227.

——— 2007a. 'Can the Retreat from Equality Be Reversed? An Assessment of Redistributive Fiscal Policies from Deng Xiaoping to Wen Jiabao'. In *Paying for Progress in China: Public Finance, Human Welfare and Changing Patterns of Inequality*, edited by Vivienne Shue and Christine Wong. Routledge Contemporary China Series 21, 12–28. London: Routledge.

——— 2007b. 'Budget Reform in China'. *OECD Journal on Budgeting* 7 (1): 1–24. doi: 10.1787/budget-v7-art2-en.

——— 2009. 'Rebuilding Government for the 21st Century: Can China Incrementally Reform the Public Sector?' *The China Quarterly* 200 (December): 929–52. doi: 10.1017/S0305741009990567.

——— 2010. 'Paying for the Harmonious Society'. *China Economic Quarterly* 14 (2): 20–25.

———— 2012. 'Performance, Monitoring, and Evaluation in China'. PREM Notes No. 23. The World Bank. https://openknowledge.worldbank.org/handle/10986/17083.

———— 2013a. 'Paying for Urbanization: Challenges for China's Municipal Finance in the 21st Century'. In *Financing Metropolitan Governments in Developing Countries*, edited by Roy W. Bahl, Johannes F. Linn, and Deborah L. Wetzel. Cambridge, MA: Lincoln Institute of Land Policy.

———— 2013b. 'Reforming China's Public Finances for Long-Term Growth'. In *China: A New Model for Growth and Development*, edited by Ross Garnaut, Cai Fang, and Ligang Song, 1st ed., 199–219. ANU Press. https://doi.org/10.22459/CNMGD.07.2013.10.

———— 2016. 'Budget Reform in China: Progress and Prospects in the Xi Jinping Era'. *OECD Journal on Budgeting* 15 (3): 27–36. doi: 10.1787/budget-15-5jm0zbtm3pzn.

———— 2017. 'The Financial Crisis and the Challenge of Fiscal Federalism in China: The 2008 Stimulus and the Limits of China's Intergovernmental System'. In *The Future of Federalism*, edited by Richard Eccleston and Rick Krever, 249–270. Cheltenham, UK; Northampton, MA, USA: Edward Elgar Publishing. doi: 10.4337/9781784717780.00020.

———— 2018a. *ASIA 90011 – China's Economic and Social Development*. Lecture Session 6, 28 August 2018, Semester 2. University of Melbourne.

———— 2018b. 'An Update on Fiscal Reform'. In *China's 40 Years of Reform and Development: 1978–2018*, edited by Ross Garnaut, Ligang Song, and Cai Fang, 1st ed., 270–290. ANU Press.https://press.anu.edu.au/publications/series/china-update-series/china's-40-years-reform-and-development-1978-2018.

Wong, Christine, and Richard M. Bird. 2005. 'China's Fiscal System: A Work in Progress'. *Rotman School of Management Working Paper No. 07-11*. doi: 10.2139/ssrn.875416.

Wong, Christine, and Randong Yuan. 2020. 'Managing across Levels of Government: The Challenge of Pension Reform in China'. In *Ageing and Fiscal Challenges across Levels of Government, by OECD and Korea Institute of Public Finance*, edited by Junghum Kim and Sean Dougherty, 77–100. OECD Fiscal Federalism Studies. OECD. doi: 10.1787/8e800011-en.

World Bank. 1997. *Old Age Security: Pension Reform in China*. Washington, DC: The World Bank.

————. 2002. *China – National Development and Sub-National Finance: A Review of Provincial Expenditures*. The World Bank. https://ideas.repec.org/p/wbk/wboper/15423.html.

————. 2007. *China: Public Services for Building the New Socialist Countryside*. The World Bank. https://ideas.repec.org/p/wbk/wboper/7665.html.

————. 2016. *Life Expectancy at Birth, Total (Years)*. http://data.worldbank.org/indicator/SP.DYN.LE00.IN.

————. 2018a. 'Fertility Rate, Total (Births per Woman)'. 2018. https://data.worldbank.org/indicator/SP.DYN.TFRT.IN.

————. 2018b. 'GDP Growth (Annual %)'. 2018. https://data.worldbank.org/indicator/NY.GDP.MKTP.KD.ZG.

————. 2018c. 'Life Expectancy at Birth, Total (Years)'. 2018. https://data.worldbank.org/indicator/SP.DYN.LE00.IN.

————. 2019. *Life Expectancy at Birth, Total (Years)*. http://data.worldbank.org/indicator/SP.DYN.LE00.IN.

————. 2020a. 'GDP Growth (Annual %)'. 2020. https://data.worldbank.org/indicator/NY.GDP.MKTP.KD.ZG.

————. 2020b. *Life Expectancy at Birth, Total (Years)*. https://data.worldbank.org/indicator/SP.DYN.LE00.IN.

————. 2021. 'Fertility Rate, Total (Births per Woman)'. 2021. https://data.worldbank.org/indicator/SP.DYN.TFRT.IN.

Wu, Ling. 2013. 'Inequality of Pension Arrangements Among Different Segments of the Labor Force in China'. *Journal of Aging & Social Policy* 25 (2): 181–96. doi: 10.1080/08959420.2012.735159.

Xi, Jinping. 2021. 'Xi Jinping's Speech on the 28th Group Learning Session of the Politburo of the Chinese Communist Party'. In Chinese. 26 February 2021. http://politics.people.com.cn/n1/2021/0228/c1024-32038460.html.

Xia, Boguang. 2001. 'Early Retirement: A Constant Pain for Pension Funds?'. In Chinese. *China Social Security* 5: 6–10.

Xinhua Net. 2002. 'Background Information: "Two Assurances" and "Three Guarantee Lines" to Improve Social Security'. In Chinese. *Xinhua Net* 2002. http://news.sina.com.cn/c/2002-11-11/1501804052.html.

———. 2015. 'Office of Poverty Alleviation of State Council: China's Poverty Line Has Become Higher than That of the World Bank'. In Chinese. *Xinhua Net.* 16 December 2015. http://www.xinhuanet.com/gongyi/2015-12/16/c_128535730.htm.

———. 2017a. 'A Deputy Director of Social Security Bureau in Zaozhuang Embezzled 36 Million of Pension Funds'. In Chinese. *Xinhua Net.* 2017. http://www.xinhuanet.com/2017-01/26/c_1120385470.htm.

———. 2017b. 'Clear Trend of Migrant Workers Returning to the Traditional Main Net-Exporter Provinces'. In Chinese. *Xinhua Net.* 2017. http://www.xinhuanet.com/politics/2017-02/09/c_1120440547.htm.

———. 2017c. 'Investigation on Couriers Social Insurances Situation: Low Compliance on Social Security from Franchise Networks'. In Chinese. *Xinhua Net.* 2017. http://www.xinhuanet.com//local/2017-03/12/c_1120611480.htm.

———. 2018a. '27 Million Private Companies in China Now and Contributing over 50% of Fiscal Revenues'. In Chinese. *Xinhua Net.* 2018. http://www.xinhuanet.com/politics/2018-05/01/c_1122767077.htm.

———. 2018b. 'Retirement for Sale? A Major Case of Pension Fraud in Luohe Is Solved'. In Chinese. *Xinhua Net.* 2018. http://www.xinhuanet.com/legal/2018-07/19/c_1123148418.htm.

———. 2019a. 'Beijing to Lower Pension Contribution Rates from 1 May 2019'. In Chinese. *Xinhua Net.* 2019. http://www.xinhuanet.com/2019-05/01/c_1124441931.htm.

———. 2019b. 'Several Provinces Speed up Provincial Unification of Pension Insurances'. In Chinese. *Xinhua Net.* 2019. http://www.xinhuanet.com/fortune/2019-09/02/c_1124948207.htm.

Xinhua News Agency. 2006. 'Social Insurance Funds Recovered from Misappropriation Amounting to 16 Billion Yuan in Seven Years in China'. In Chinese. *Xinhua News Agency.* 2006. http://www.gov.cn/jrzg/2006-04/24/content_263128.htm.

———. 2014. 'Chinese President Proposes Asia-Pacific Dream'. *Xinhua News Agency.* 2014. http://www.apec-china.org.cn/41/2014/11/09/3@2418.htm.

———. 2018a. 'China's Urbanisation Rate Rose to 58.52%: To Release New Momentum for Development'. In Chinese. *Xinhua News Agency*, 4 February 2018. http://www.gov.cn/guowuyuan/2018-02/04/content_5263778.htm.

———. 2018b. '*Minimum Wage in Shanghai to Be Raised to 2,420 Yuan*'. In Chinese. Website of Central Government of People's Republic of China. 21 March 2018. http://www.gov.cn/xinwen/2018-03/21/content_5276305.htm.

Xiong, Jian. 2016. 'Some Data about the Demobilisation of the Chinese People's Liberation Army'. In Chinese. *People's Daily Online.* 15 July 2016. http://politics.people.com.cn/n/2015/0903/c1001-27543904.html.

Xiong, Jianpeng, and Ge Wang. 2016. 'Embezzlement of 9 Million Pension Funds by Social Security Agency Employee in Chengde: 248 Victims Asked to Fork out for the Loss'. In Chinese. *The Paper.* 2016. http://www.thepaper.cn/newsDetail_forward_1486922.

Xu, Li, and Chun Wan. 2008. 'The Evolution of Chinese Old-Age Insurance System: 1951–2008'. In Chinese. *Reform* 178 (12): 125–131.

Yang, Yixin, and Wenjiong He. 2016. 'Can Increasing the Contributory Period Effectively Improve the Financial Sustainability of Pension Fund? Based on the Actuarial Evaluation of Pay-as-You-Go and Funded System'. In Chinese. *Population Research* 40 (3): 18–29.

Yu, Hong, and Yi Zeng. 2015. 'Retirement Age, Fertility Policy and the Sustainability of the Basic Pension Fund in China'. In Chinese. *Journal of Finance and Economics* 41 (6): 46–57.

Yuan, Lei. 2014. 'Can the Gap of Social Pension Fund Problem Be Settled by Postponing Retirement Age? Simulations of Three Postponing Retirement Age Proposals under 72 Different Conditions'. In Chinese. *Population and Economics* 4: 82–93.

Yuan, Xin, and Neng Wan. 2006. 'Is Retirement Age Delay Effective for Alleviating the Pressure from Population Ageing?' In Chinese. *Population Research* 30 (4): 47–54.

Yuan, Zhigang, Jin Feng, Jinfeng Ge, and Qin Chen. 2016. *Pension Economics: An Illustration of the Challenges Faced by China*. In Chinese. Beijing: China CITIC Press.

Yuan, Zhongmei. 2013. 'Discussion of Delay Retirement and Pension Replacement Rate'. In Chinese. *Population and Economics* 1: 101–6.

Zeng, Yi. 2011. 'Effects of Demographic and Retirement-Age Policies on Future Pension Deficits, with an Application to China'. *Population and Development Review* 37 (3): 553–69.

Zhang, Li, and Meng Li. 2018. 'Acquired but Unvested Welfare Rights: Migration and Entitlement Barriers in Reform-Era China'. *The China Quarterly* 235 (September): 669–92. doi: 10.1017/S030574101800084X.

Zhang, Li, Rhonda Vonshay Sharpe, Shi Li, and William A. Darity. 2016. 'Wage Differentials between Urban and Rural-Urban Migrant Workers in China'. *China Economic Review* 41 (December): 222–33. doi: 10.1016/j.chieco.2016.10.004.

Zhao, Qing, Xiaojun Wang, and Haijie Mi. 2015. 'Multidimensional Evaluation of the Adequacy of Pension System in China'. In Chinese. *Statistical Research* 32 (6): 36–41.

Zheng, Bingwen. 2012. 'China: An Innovative Hybrid Pension Design Proposal'. In *Nonfinancial Defined Contribution Pension Schemes in a Changing Pension World*, edited by Robert Holzmann and Edward Palmer, 189–212. The World Bank. doi: 10.1596/9780821388488_CH06.

———— 2016. *China Pension Report 2016*. In Chinese. Beijing: Economic Management Press.

Zheng, Yongnian. 2015. '*Prevent the "New Normal" from Becoming the "Abnormal"*'. In Chinese. *China Reform Forum*. 2015. http://people.chinareform.org.cn/Z/zyn/Article/201504/t20150429_224070.htm.

Zhengyi Net. 2019. 'Investigation on Delivery Personnel's Labour Rights: No Contract, No Social Security, but with Cash Fines'. In Chinese. *Zhengyi Net*. 2019. http://news.jcrb.com/jxsw/201904/t20190423_1992952.html.

Zhou, Li-an. 2010. *Incentives and Governance: China's Local Governments*. Singapore: Cengage Learning Asia Pte Ltd.

Zhu, Huoyun, and Alan Walker. 2018. 'Pension System Reform in China: Who Gets What Pensions?' *Social Policy & Administration* 52 (7): 1410–24. doi: 10.1111/spol.12368.

Zhu, Li. 2003. 'Characteristics and Social Status of the Peasant Worker Class'. In Chinese. *Journal of Nanjing University (Philosophy, Humanities and Social Sciences)* 40 (6): 41–50.

Zhu, Qingfang. 1998. 'The Characteristics, Causes and Solutions for Urban Poverty in China'. In Chinese. *Social Sciences Research* (1): 62–66.

Zhu, Yukun. 2002. 'Recent Developments in China's Social Security Reforms'. *International Social Security Review* 55 (4): 39–54.

Appendix A
Interview guides

Guide for interviews with local officials for the fieldwork in 2017

Theme 1. Enquire about interviewee's observations of the development of the pension system in the local county or city and its outcome

Q: What is your reflection on the development of the pension system in your county or city over the past decade? What are the particular changes worth noting that you have observed over the time?

Q: How would you rate the success of the policy implementation since you have been involved in the provision of pension services in your county or city?

Q: What are the differences that you see in the development of the pension system in your county or city in comparison to other regions in China?

Theme 2. Enquire about interviewee's opinion of or experience in the financial management of the pension system in the local county or city

Q: In your role as a government official, have you been involved in the financial management of the pension system in your county or city?

Q: To your knowledge, is there a surplus or deficit in the current running of the pension system in your county or city? How is the surplus, if any, managed? How is the deficit, if any, financed?

Q: To your knowledge, how reliant is the financing for pension expenditures on transfers from within the province? How reliant is the financing for pension expenditures on transfers from the central government?

Theme 3. Enquire about interviewee's observations on the three pension schemes, namely the UES, the BRS and the PES

Q: What is your view on the three pension schemes that constitute China's current pension system? How have they each played out in your county or city?

Q: How would you describe the relationship among these three schemes? What changes have you noticed of such relationship since you became involved in the management of the local pension system?

Q: What is your perspective on the future development of these three schemes, individually and collectively?

Theme 4. Enquire about interviewee's views on the effect of the pension system on the enterprises and investment environment in the local county or city

Q: To your knowledge, how is the pension system affecting state-owned enterprises in your county or city? How is the pension system affecting private enterprises in your county or city?

Q: To your knowledge, how is the pension system affecting inward investment in your county or city?

Theme 5. Enquire about interviewee's views on the role of the central government and the role of the local government in the provision of pension services in the local county or city

Q: What role do you think the central government should play in the overall development of the pension system? What role do you think the local government should play in the pension system?

Q: Should the running of the pension system be centralised at the national level? Would the pension system be more efficient and equitable if it were run by the central government?

Theme 6. Enquire about interviewee's views on the existing barriers and emerging challenges to policy delivery in local context

Q: What is your view on the fundamental barriers or constraints of policy implementation related to pension reform?

Q: What is your view on the upcoming challenges to pension reform in the future? How is the change in the population structure in your county or city affecting the pension system?

Q: If you could make one and only one change to the existing pension system in China, what would that be?

Guide for interviews with local officials for the fieldwork in 2019

Theme 1. Enquire about the financing mechanism of pension schemes

Q: How are the three pension schemes, namely the UES, the BRS and the PES, financed, respectively, in your county or city?

Q: For each pension scheme, is there deficit or surplus in the running of the local pension funds?

Q: For any pension scheme with deficit, how is the deficit financed?

Q: If the deficit is financed or at least partly financed by transfer payment from higher level government, how is the amount of transfer payment determined every year?

Q: Are there any rules for determining the amount of such transfer payment? Are there any preconditions for the local government to meet before obtaining such transfer payment from higher level government?

Q: Does the local government need to negotiate with the higher level government on the amount of the transfer payment on a yearly basis?

Q: Does the local government self-finance part of the deficit? If so, how is it done?

Q: For any pension scheme with surplus, how is the surplus managed financially?

Q: How many options does the local government have for investing the surplus, e.g. short-term bank deposits or treasury bonds? If the surplus is saved as bank deposits, does the local government have the right to determine which bank to save the surplus with? Which banks are involved?

Q: What is the level for the pooling of pension funds for each pension scheme, e.g. at county/district, prefectural or provincial level?

Theme 2. Enquire about the collection of pension contributions

Q: Has the local tax office taken over the collection of pension contributions? If so, does the tax office charge for handling fees for the collection of pension contributions? If so, how much are the handling fees?

Q: For firms with pension contributions in arrears, how are they dealt with? Which government department is in charge of such issues, e.g. by the bureau of human resources and social security or the tax office?

Q: How are pension contributions collected? Do companies report to the tax office or the bureau of human resources and social security by themselves? If so, do the numbers reported by the companies tally with the numbers used for company income tax calculation and the numbers used for personal income tax calculation for the employees?

Q: Can companies still report the wages for their employees at the lower limit of wages allowed for calculating pension contributions, e.g. at 60% of the local average wage?

Q: Are cash transactions involved in the collection of pension contributions for any pension scheme?

Q: For the UES, what are the lower limit and higher limit of the wages that can be used to calculate pension contributions?

Q: Are all the firms facing the same lower and higher limits of wages? Or are different types of firms treated differently, e.g. SOEs versus small private companies?

Q: Do companies in industry parks receive any discounts in pension contribution rates or can they report lower-than-normal wages for computing pension contributions, e.g. 40% of local average wage?

Q: How are buy-in fees handled? Is the tax office in charge of the collection of buy-in fees now? Who are eligible for the buy-in option? Which government department, e.g. the bureau of human resources and social security or some other government department, determines the eligibility for the buy-in option? According to which policies or legal documents? Are there any local regulations on determining the eligibility of buy-in option for individuals?

Q: Is the prior employment history in the collective enterprises before the 1997 pension reform recognised as pension contribution history for the former employees in the collective enterprises? Are there any other groups of people for whom the prior employment or service (e.g. for former military personnel) history is recognised as pension contribution history? How are the costs for such recognition financed? Is there any transfer payment from higher level government earmarked for financing such costs?

Q: Are there any issues with coordination between the bureau of human resources and social security and the tax office?

Theme 3. Enquire about the determination of pension benefits and local average wage

Q: How are pension benefits determined for each pension scheme?

Q: How is the local average wage defined and calculated? Which groups of companies and industries are included for calculating the local average wage? Are there any groups of workers in the local labour force excluded from the computation of the local average wage?

Q: Which government agency is in charge of determining the local average wage?

Q: Is the local average wage used for calculating the pension benefits computed at county/district, prefectural or provincial level?

Q: Is the local average wage used for calculating pension benefits the same as the one used for calculating pension contributions? If they are not the same, what is the difference?

Appendix B

Calculation for the actuarial present value of buy-in option for the UES

The pension benefits for a retiree of the UES consist of two parts, i.e. a regular monthly retirement income paid while the retiree is alive, and a death benefit paid upon the death of the retiree. The former is a life annuity, while the latter is a whole life insurance. The actuarial present value (APV) for the buy-in option thus also consists of two parts that correspond to the APV of the life annuity and the whole life insurance, respectively. For simplicity, we assume that pension benefits are paid annually at the end of the year and that the death benefit is paid at the end of the year of death.

Therefore

$$\text{APV}\left(\text{BI}\right) = \text{AB} \times a_x + \text{DB} \times A_x$$

where APV(BI) denotes the APV of the buy-in option, AB denotes the annual pension benefits, a_x denotes the APV for one unit of life annuity issued to a person aged x, DB denotes the death benefit and A_x denotes the APV for one unit of whole life insurance issued to a person aged x.

Following the standard actuarial techniques for valuing life annuity and whole life insurance, we have:

$$a_x = \sum_{t=1}^{\infty} \left(1+r\right)^{-t} {}_t p_x$$

where t denotes the number of the years a person lives into the future, r denotes the discount rate, and ${}_t p_x$ denotes the probability that a person aged x survives to age $x + t$.

We also have:

$$A_x = \sum_{t=0}^{\infty} \left(1+r\right)^{-(t+1)} {}_t p_x q_{x+t}$$

where q_{x+t} denotes the probability that a person aged $(x + t)$ dies within one year.

When calculating the APV of the buy-in option for females, $x = 55$. When calculating the APV of the buy-in option for males, $x = 60$. This is due to the difference in retirement age. The gender difference is also manifested in mortality rates, which leads to different values for a_x and A_x between females and males.

The data used for mortality rates are obtained from *China Life Insurance Mortality Table (2010-2013)* released by the China Insurance Regulatory Commission. The discount rate used is 3%, which is based on the risk-free rate estimated from the latest yield curve of the Chinese government bonds. The annual pension benefits are assumed to grow at 5% per annum, which is a conservative estimate given that the average annual growth rate for the national average pension benefits was 10.2% in the past 15 years from 2004 to 2019 and at 8.7% in the past 10 years from 2009 to 2019. If the government is committed to maintain the growth above 5% per annum in the future, the APV will be higher than those estimated here. Other data required for the calculation, such as the annual pension benefits and the death benefit, were obtained from the fieldwork interview in County X in July 2019.

Index

Page numbers followed by n indicate notes.

Printed in the United States
by Baker & Taylor Publisher Services

Printed in the United States
by Baker & Taylor Publisher Services